Moral Dilemmas

Moral Dilemmas

Edited by

Christopher W. Gowans

New York Oxford
OXFORD UNIVERSITY PRESS
1987

Oxford University Press

Oxford New York Toronto
Delhi Bombay Calcutta Madras Karachi
Petaling Jaya Singapore Hong Kong Tokyo
Nairobi Dar es Salaam Cape Town
Melbourne Auckland

and associated companies in
Beirut Berlin Ibadan Nicosia

Published by Oxford University Press, Inc.,
200 Madison Avenue, New York, New York 10016

Oxford is a registered trademark of Oxford University Press

Library of Congress Cataloging-in-Publication Data
Moral dilemmas.
Bibliography: p.
1. Ethics. 2. Dilemma. I. Gowans, Christopher W.
BJ1012.M6315 1987   170   86-16320
ISBN 0-19-504272-7
ISBN 0-19-504271-9 (pbk.)

9 8 7 6 5 4 3 2 1

Printed in the United States of America

To
Milt Yanicks
my first philosophy teacher

Preface

The readings collected here pertain in various ways to the question of whether there can be moral dilemmas. To many, this may seem an odd topic for debate insofar as it seems obvious that we are often perplexed about what morality expects of us. However, what is at issue here is not the evident fact of occasional moral perplexity but whether it can actually be the case that a person morally ought to do one thing and morally ought to do another when both cannot be done. It has seemed to most philosophers that such a situation could not obtain (except perhaps as a result of previous wrongdoing). But this viewpoint has been challenged by several analytic philosophers in recent years, most forcefully by Williams, and the ensuing debate is the topic of this collection.

The first four selections represent what I take to be the main historical traditions from which the contemporary debate has developed: those of Kant, the utilitarians, Hegel, and the British intuitionists. Each of these readings has been edited from a book (in the case of Mill's selection, two books); the reader should be aware that these selections are taken out of context and that some sections have been omitted. The remaining selections have all been written in the last twenty-five years. With two exceptions—those by Nagel and

Hare—they were originally published in professional philosophy journals. Nagel's paper is taken from his collection of essays, *Mortal Questions,* and appears here without revision. The selection from Hare consists of chapter 2 and all but the last section of chapter 3 of his book, *Moral Thinking.* The other essays are reprinted here as they first appeared, without alteration or abridgment, with the exception of Lemmon's article, from which a section on akrasia and bad faith has been deleted.

On occasion, where principles of deontic logic are at issue, there is some use of formal logic in these essays, primarily in those by van Fraassen and McConnell. This may be an obstacle to those unfamiliar with the notation. However, in a classroom at least, this should not be an insurmountable problem, as the notation can be explained easily and the formal discussion of deontic principles, though significant, is only one facet of the debate as a whole.

The bibliography includes works in which there is discussion of moral dilemma or moral conflict, as well as of related issues raised in the debate. Most, but by no means all, of these works have been written by analytic philosophers in the last twenty-five years. The bibliography also includes philosophical works I have referred to in my introduction; however, it includes almost nothing on applied ethics, for which other bibliographies are available.

I would like to express my gratitude to those who have granted permission to reprint the selections in this volume. I would also like to thank the many people who have been helpful to me in the preparation of this book, in particular the anonymous readers for Oxford University Press; my editors at Oxford, Cynthia Read and Marion Osmun; the students in my graduate class at Fordham University in the fall of 1984; and also Brian Leftow, Mike Morris, Dave Solomon, Peter Vallentyne, Michael Zimmerman, and especially Margaret Walker, with whom I have had many illuminating discussions and who has generously commented on various aspects of this anthology throughout its development. I am also grateful to Victoria Burke, who was expecially helpful in assisting me with the review of the proofs. Finally, I would like to express my deep appreciation to Anne, whose patience and unfaltering support were invaluable as I progressed on this work over the past two years.

New York   C.W.G.
November 1986

Contents

Moral Dilemmas

INTRODUCTION

The Debate on Moral Dilemmas

Christopher W. Gowans

1

A moral dilemma is a situation in which an agent S morally ought to do A and morally ought to do B but cannot do both, either because B is just not-doing-A or because some contingent feature of the world prevents doing both. That there are cases of *apparent* moral dilemma can hardly be denied. Contemporary work in applied ethics has shown, if nothing else, that compelling arguments can be given for incompatible positions on a variety of topics; abortion, euthanasia, capital punishment, preferential treatment, and censorship are but a few examples. MacIntyre has recently argued that this results from something unique in our culture, namely that we have inherited mere fragments of heterogeneous moral theories, each severed from the larger social context that gave it purpose (1981, chapter 2).* But this explanation, though not without merit, is surely too simple; and this is implicitly acknowledged by Mac-

---

*Works included in this volume are indicated by author and chapter or page number of this volume; references to all other works are indicated by author, year of publication, and, when appropriate, chapter or page number (for full documentation see the bibliography).

Intyre himself in his endorsement of the "Sophoclean" view that "there *is* an objective moral order, but our perceptions of it are such that we cannot bring rival moral truths into complete harmony with each other" (1981, p. 134).

In any case, we need not restrict ourselves to contemporary debates in order to see the centrality of moral conflict to human life. Throughout history, moral conflicts have been a topic not only of philosophical discussion (e.g., Kant, 1909; Plato, 1974, 331c; Sartre, 1975, pp. 354–57); they have also been, perhaps more importantly, the substance of dramatic literature (see Nussbaum, 1985, 1986). Instances abound. In Sophocles's *Antigone,* Creon declares the burial of Antigone's brother Polyneices illegal on the not unreasonable grounds that he was a traitor to the city and that his burial would mock the loyalists who defended the city, thereby threatening civil disorder; at the same time, there is reason for Creon to respect the religious and familial obligation of Antigone to bury her brother. Just as familiar and no less striking is the dilemma addressed in the speech of Brutus, in Shakespeare's *Julius Caesar,* in defense of the slaying of Caesar, his friend but ambitious leader, "not that I loved Caesar less, but that I loved Rome more" (act 3, scene 2). More recent and closer to home is the choice of Nora, in Ibsen's *A Doll House,* between duties to husband and children and "other duties equally sacred," namely, duties to herself (act 3, Fjelde translation), and also the choice of Barbara, in Shaw's *Major Barbara,* between discontinuing her efforts on behalf of the bodily and spiritual salvation of the poor and accepting donations that have their origin in profits of a liquor and a munitions manufacturer.[1]

The substance of drama is the substance of life. These plays are of continuing consequence because they portray conflicts such as each of us confront potentially if not actually in our own lives. And yet, with few exceptions, philosophers from Plato on have viewed moral dilemmas as mere appearances. This has certainly been the case in the two predominant traditions of modern moral philosophy—Kantianism and utilitarianism. Both Kantians and utilitarians have thought that, for any apparent conflict, either one of the conflicting ought statements is not true or the two statements do not really enjoin incompatible actions. It is admitted that there are cases in which there may be uncertainty about the resolution of an apparent dilemma. But it is thought to be impossible that morality could actually impose upon an agent two 'oughts' when both cannot be fulfilled.

However, in recent years a number of moral philosophers in

the analytic tradition have challenged this view by suggesting that there are genuine moral dilemmas. In the process, they have challenged the common assumptions of much contemporary moral theory, both Kantian and utilitarian. Indeed, this is but a part of the broader challenge to these two orientations that has emerged in recent years (e.g., Blum, 1980; MacIntyre, 1981; Williams, 1973b, 1985). It is hardly surprising, then, that this challenge has met with substantial opposition. The purpose of this anthology is to bring together in a single volume the principal contributions to this debate. In this introductory essay, the central themes and arguments that have emerged from these discussions will be sketched and evaluated. Sections 2 through 6 concern those philosophical traditions out of which the contemporary discussion develops; sections 7 through 15 approach this discussion in systematic fashion.

2

Though the current debate takes its bearings mainly by reference to the Kantian and utilitarian traditions (as well as to alternative modern approaches, most notably intuitionism), it is worth noting briefly that discussion of moral conflict has roots deep in the history of moral philosophy. Among classical theories, the doctrine of the unity of the virtues in Aristotle (1980, VI, 13) implies that there can be no conflict among the virtues; and the natural law doctrine of Aquinas specifically precludes moral dilemmas.[2] However, other classical approaches at least open the possibility of moral dilemmas, in particular command theories (discussed by van Fraassen, chapter 7 and Donagan, chapter 14), such as the divine command theory (see Quinn, 1978, p. 38, and 1979, pp. 319–21) and contract theories that justify moral principles on the basis of an agreement among persons (see Harman, 1975b, and Plato, 1974, 358e–59b). In general, to put the matter in traditional terms, it might be supposed that while rationalist theories (those that found morality on reason) preclude moral dilemmas, voluntarist theories (those that ground morality on expressions of will, divine or human) do not preclude moral dilemmas. However, this dichotomy is misleading. For one thing, a voluntarist theory may put rational constraints on legitimate expressions of will (as in Rawls, 1971). Moreover, it is not evident, at least without further specification, that reason does preclude moral dilemmas: by itself, a dilemma is not inconsistent (see sections 10 and 11 below).

3

In modern philosophy, Kant provides one of the standard arguments against the possibility of moral dilemmas. In the selection in this volume (chapter 1), he declares unequivocally that "a *conflict of duties* and obligations is inconceivable" (p. 39) on the grounds that the rules expressing moral duty declare certain actions to be "necessary" and that two rules declaring actions necessary cannot conflict. Thus, if it is a duty, and hence a moral necessity, that a person do *A*, then it cannot also be a duty, and hence a moral necessity, that the person do something incompatible with *A*. Kant acknowledges that there can be conflicting "*grounds* of obligation." But in such a case one of the grounds is not a sufficient ground. Hence, there are not two conflicting obligations, one of which prevails, but two conflicting grounds of obligation, one of which prevails; the result is that there is only one actual obligation.

This brief argument has been widely discussed and variously interpreted.[3] One source of puzzlement is Kant's admission that, though there are no conflicting obligations, there are conflicting grounds of obligation. This raises questions about both the status of these grounds and the basis for resolving conflicts among them. More to the point, however, is Kant's rejection of genuine conflicts of obligation. Donagan, who develops a Kantian position (chapter 14), finds this argument not only clear but convincing. However, others, including several authors in this volume, find it unacceptable. A major point of contention, about which there is more below (section 7), is the claim that in apparent conflicts of obligation there can be only one actual obligation, the other exerting no residual influence.

It is clear, in any case, that the central premise in Kant's argument is the claim that some actions are morally necessary. For Kant, moral rules (or laws) are unconditional imperatives that declare certain actions that agents are free to perform, or not, either morally necessary or morally impossible, meaning that from the point of view of practical reason the agent either must or must not perform them. Actions that are neither morally necessary nor morally impossible are "morally indifferent" or "permissible." These three categories—the necessary, the impossible, and the permissible—are taken to be exclusive and exhaustive: every action falls into one and only one. Hence, it is incoherent to suppose that an action could be both necessary and impossible or that two actions could both be necessary when doing one precludes doing the other.

As will be seen below (section 11), the concept of moral necessity underlies a major class of objections to moral dilemma, those based on standard principles of deontic logic. However, in view of Kant's distinction between perfect and imperfect duties, it appears that Kant's own position is somewhat more complicated than suggested thus far. Perfect (or narrow) duties, according to Kant, prescribe or prohibit all instances of specific kinds of action. For this reason, there is no latitude in deciding how to fulfill these duties. But imperfect (or wide) duties are different. These duties prescribe not every instance of a specific kind of action but an unspecific pursuit of ends. And these duties, Kant says, "cannot specify precisely what and how much one's actions should do toward the obligatory end" (p. 45). Kant cautions that this "is not to be taken as a permission to make exceptions" to these maxims; rather, it indicates "a permission to limit one maxim of duty by another" (pp. 45–46).

There has been much debate about the distinction between perfect and imperfect duties.[4] However, it seems clear that in the argument against conflicts of duties, Kant plainly means that it is perfect duties that cannot conflict, because they make all instances of specific kinds of action necessary (but see Donagan, chapter 14). Since imperfect duties do not make any particular action necessary, there is no reason to deny, and in fact it is clearly the case, that these duties do conflict in the sense that, on a particular occasion, pursuing one end (e.g., the happiness of my neighbor) may mean not pursuing another (e.g., the development of my talent). When this happens, the duties themselves prescribe no specific course. The agent may act as he pleases, as long as he does not abandon an imperfect duty altogether, and as long as his action violates no perfect duty. Of course, imperfect duties may not conflict in the sense that pursuing the end of one *to any extent* precludes pursuing the end of another.[5]

4

Though there are significant differences between the Kantian approach and the utilitarian approach, they agree in their rejection of the possibility of genuine moral dilemmas. However, for classical utilitarians such as Mill, it is not so much argued as assumed that apparent dilemmas cannot be genuine, and it is then claimed that one advantage of utilitarianism is that it provides the means for their resolution. Thus, in the selection here (chapter 2), Mill acknowledges that "unequivocal cases of conflicting obligation" do arise but goes on to declare, "If utility is the ultimate source of moral obligations, utility may be invoked to decide between them when their demands are incompatible" (pp. 54–55). And while he grants that it may not be easy to apply the standard of utility, he argues that "in other systems, the moral laws all claiming independent authority, there is no common umpire entitled to interfere between them" (p. 55).[6]

Two questions are raised by this claim: whether utilitarianism can resolve apparent conflicts, and whether other theories cannot. With regard to the second question, Mill claims that there can be only one standard of value. "If there were several ultimate principles of conduct," he argues, "the same conduct might be approved by one of those principles and condemned by another; and there would be needed some more general principle, as umpire between them" (p. 52). The implication is that the more general, umpire principle would become the sole ultimate principle; hence, there would no longer be several ultimate principles. For Mill goes on to speak of a single principle "with which all other rules of conduct were required to be consistent, and from which by ultimate consequence they could all be deduced" (p. 53).

The argument, thus formulated, is an argument for a monist theory, and not for utilitarianism per se (cf., Gewirth, 1978, p. 12). But it is clear that Mill regards utilitarianism as the main alternative to the pluralism of the intuitionist theories he is opposing. In any case, the argument is unconvincing. Even if two principles, *P1* and *P2*, conflict, it does not follow that an "umpire principle" for resolving the conflict becomes the sole ultimate moral principle. For, from the principle (say) '*P1* takes priority over *P2*' neither *P1* nor *P2* can be deduced. A priority principle concerns the comparative importance of first-order principles; it says nothing about their absolute importance. Hence, a pluralist can perfectly well accept *P1, P2,* and '*P1* takes priority over *P2*' as ultimate principles: no one need entail the others.

The other question is whether utilitarianism itself provides a means for resolving conflicts. For utilitarianism, rightness is a function of the maximization of utility (variously interpreted as pleasure, happiness, goodness, preference-satisfaction, etc.); as applied directly to actions, it declares an action *A* to be right if and only if the consequences of *A* have greater utility than the consequences of possible alternative actions. The intuitive plausibility of the argument that utilitarianism can resolve all apparent moral conflicts rests upon the assumption that utility, however understood, is a homogeneous value to which all morally relevant considerations may be reduced. However, it has proved notoriously difficult to defend such a concept of utility (see section 8). It is possible, of course, for the utilitarian to regard utility as a composite of heterogeneous values (see Sen, 1980–81), but if this avenue is followed, the alleged advantage of utilitarianism vis-à-vis conflict resolution is at least diminished. Hence, utilitarianism is faced with a dilemma: *either* it defends the homogeneity of utility, in which case it may succumb to an implausible reductionism, *or* it admits the heterogeneity of utility, in which case it may forego its purported advantage in resolving conflicts.

In the current discussion, the utilitarian viewpoint is represented by Hare (chapter 11). His approach to moral conflict depends on the distinction between "intuitive" and "critical" levels of moral thinking. The critical level is the "epistemologically prior" one; at this level specific actions are judged right or wrong from a utilitarian viewpoint. On the basis of these judgments, Hare supposes, a set of relatively simple (that is, unspecific) *prima facie* principles may be developed for the purposes of moral education and everyday guidance. It is these principles that we ordinarily rely on in our intuitive moral thinking; and in most cases, if they are properly formulated, they will give correct and unequivocal guidance. However, in unusual cases they may conflict. The proper response to this, Hare argues, cannot be made at the intuitive level of *prima facie* principles. Rather, there must be a return to the critical level. That is, we must determine the proper resolution of the conflict on the basis of a utilitarian analysis of the particular situation. Thus, Hare writes, at the intuitive level "conflicts are indeed irresolvable; but at the critical level there is a requirement that we resolve the conflict" (p. 206; cf., Primorac, 1985). Hare's argument has the advantage of explaining why it can plausibly appear that there are genuine moral dilemmas even if in fact there are none; as will be seen below (section 7), this allows him to respond to one important class of arguments for dilemma.[7]

5

A common feature of most contemporary forms of both Kantian
and utilitarian theories, if less clearly of Kant and Mill themselves,
is monism: each theory attempts to reduce moral considerations to
a single dimension and thereby attempts to eliminate all apparent
moral conflicts. However, there are other traditions in modern philosophy that allow more diversity of moral consideration and hence
a more significant role for moral conflict, even if they do not, in the
final analysis, allow genuine moral dilemmas. One of these traditions has its roots in Hegel, the other in the seventeenth and eighteenth century British intuitionists.[8]

Hegel's view of moral conflict, though rarely his main object of
attention, is implicit in many discussions, among them his account
of tragedy (1975, volume 2, pp. 1194–99; 1977, pp. 443–50). What
tragedy requires, Hegel thinks, is that the aims of two characters be
such that each character, "if taken by itself, has *justification*," while
at the same time each can achieve his aim "only by denying and
infringing the equally justified power of the other" (1975, volume 2,
p. 1196). But tragedy also requires a resolution of this conflict
through the exercise of "eternal justice." The significant feature of
this resolution, however, is that it involves a reconciliation of the
conflicting aims rather than a complete renunciation of one in favor
of the other. It achieves harmony only by overriding "the relative
justification of one-sided aims and passions" (1975, volume 2, p.
1198; cf., Scheler, 1963). What must be forthcoming from the Hegelian position, of course, is an account of *how* partially justified but
conflicting aims may be reconciled. To postulate reconciliation
either at the end of history or from the perspective of "eternal justice," however appealing, does little by itself to constructively confront *our* conflicts.

A view with distinct Hegelian roots is offered in the selection
by Bradley (chapter 3). Bradley's account of moral conflict begins
with the charge, derived from Hegel, that the Kantian concept of
"duty for duty's sake" is purely formal and so without content. As
soon as we move from this formal level to particular duties, he says,
it becomes clear that "collision of duties" is quite common. In fact,
"*every* act can be taken to involve such collision" (p. 63). True, we
ordinarily think that moral laws are inviolable, but reflection shows
that there are no laws that are not to be broken in some circumstance. For example, when Kant (1909) claims that lying is always

wrong, he fails to see that there are "duties above truth-speaking" (p. 63).

The inevitability of moral conflict is a consequence of the central thesis in Bradley's moral philosophy, also derived from Hegel, that the ultimate good for human beings is "self-realization." Bradley sees the realization of the self as occurring in three spheres. The first is that of "my station and its duties" (cf., Lemmon, chapter 5, and van Fraassen, chapter 7). For Bradley, a person's identity is connected essentially to the community of which one is a part. Hence, the realization of oneself is a matter of fulfilling one's social role and the duties imposed by it. Bradley does not shy away from the relativistic implication of this position— "the morality of one time is not that of another time" (1927, p. 189)—but he acknowledges that there are limitations to the morality of "my station and its duties," not the least of which is that one may live in a corrupt society. And, what is not always noticed, he goes on to argue that there are two additional modes of self-realization. The first of these involves the attainment of ideals of social relation "beyond what the world expects" (p. 73), and the second consists of "the realization for myself of truth and beauty" (p. 75).

The heart of Bradley's objection to the Kantian view is his claim that there are different moral values that sometimes enjoin conflicting actions. For Bradley, there can be conflicts both among the three spheres of self-realization and within each of them. However, he allows that when conflicts occur, one duty may be "higher" than the other, thereby suggesting that there may be a resolution of moral conflicts. But it is unclear whether such resolution leaves both duties fully in force, with one somehow taking precedence over the other, or whether it renders the lower duty inapplicable, and hence eliminates the conflict (see section 8). It is unclear, in short, whether Bradley means to claim that there are genuine or only apparent dilemmas.

Regardless, with regard to our knowledge of the resolution of conflicts, Bradley unequivocally rejects the idea that there is a process of "discursive" reasoning or "reflective deduction" that takes us from general rule to final practical conclusion. Against this, he suggests, with the intuitionists, that moral judgment is a matter of "intuition" or the "$\alpha'\iota\sigma\theta\eta\sigma\iota\varsigma$ of the $\phi\rho\acute{o}\nu\iota\mu\sigma\varsigma$" (p. 67). However, Bradley does little to elaborate on this idea, and as it stands it is manifestly inadequate. It offers no basis for distinguishing veridical from illusory intuitions and hence cannot be said to be a method of determining the correct resolution at all. In the end, Bradley declares morality to be a "self-contradiction" that must be transcended by religion (1927, pp. 313–14). Whether this can offer a constructive basis for confronting practical conflicts is another question.

6

The other modern tradition that has stressed moral diversity and conflict is that of the British intuitionists. Especially significant here is the eighteenth century philosopher Richard Price. Against the view that "the *whole* of virtue consists in BENEVOLENCE" (1969, p. 176), Price claims that there are six different "heads" of virtue, each of them "self-evident" (p. 187). And though he thinks it is "the same eternal reason that commands in them all," he acknowledges that sometimes "they lead us in contrary ways" (p. 186). When this happens, we may "be rendered entirely incapable of determining what we ought to choose" (p. 186). At best, in a case of conflict, the relevant heads of virtue can only be "weighed, if we would form a true judgment concerning it" (p. 188). However, Price gives no indication of what weighing involves.

The intuitionist view is represented here in the selection from Ross (chapter 4). Ross's position is remarkably similar to Price's (see Hudson, 1970, pp. 98–99), but he goes beyond Price in accounting for moral conflict by distinguishing two senses of duty—*prima facie* duty and duty *sans phrase*. The only duties that can conflict, Ross says, are *prima facie* duties. A *prima facie* duty is neither an actual duty nor the mere appearance of an actual duty. Rather, it is an "objective fact" about an act, specifically "the characteristic . . . which an act has, in virtue . . . of being an act which would be a duty proper if it were not at the same time of another kind which is morally significant" (p. 86). Later he says that a *prima facie* duty is a "parti-resultant attribute," that is, one belonging to an act in virtue of but one of its respects. An actual duty, by contrast, is a "toti-resultant attribute," one belonging to an act in virtue of "its whole nature" (p. 94).

Ross tentatively lists seven categories of *prima facie* duties, namely fidelity, reparation, gratitude, justice, beneficence, self-improvement, and nonmaleficence. That each of these is a *prima facie* duty, he says, is "self-evident." However, when it comes to knowledge of our actual duty in a particular situation, Ross claims, we have at best only probable opinion. What is required is a judgment about which one of the *prima facie* duties has the greatest "stringency." Such a judgment is not self-evident and cannot be reached via a deduction from the *prima facie* duties and some general rule. There are rules of thumb, perhaps, but ultimately, Ross says quoting Aristotle, "the decision rests with perception" (p. 99; cf., Broad, 1930, pp. 222–23).

Ross's doctrine of *prima facie* duties differs in an important respect from Hare's doctrine of *prima facie* principles (chapter 11). For Hare, *prima facie* principles are everyday surrogates for an ultimate moral principle, utilitarianism, from which their authority derives. Hence, when they conflict, authority reverts to the ultimate principle, from which the conflict may be resolved. For Ross, on the other hand, *prima facie* duties are not surrogates for an ultimate principle: each has independent authority in the sense that no one can be derived from another or from a higher principle. Thus, when *prima facie* duties conflict, there can be no appeal to an ultimate principle. The only appeal can be to a judgment about which duty, in the case at hand, is more stringent. Despite this, Hare and Ross agree that when *prima facies* duties conflict there *is* a resolution. For both, there are no conflicts between actual duties, and so there are no genuine dilemmas.

It is important to recognize that the intuitionism of Price and Ross contains several distinct theses, specifically (1) that moral judgments rest upon something self-evident, (2) that there is a plurality of moral principles, (3) that moral principles can conflict, and—for Ross at least—(4) that determination of the correct resolution of conflicts cannot be governed by rules but rather requires an immediate and uncertain perception of the relative weight of the conflicting principles in a specific case. The first of these theses, a form of foundationalism, is clearly independent of the others (see Sen, 1985, p. 176), and indeed, nowadays, the term 'intuitionism' is often used to refer only to the last three theses (e.g., by Raphael, 1974–75, p. 2; Rawls, 1971, p. 34; and Urmson, 1974–75, p. 111).

It is also important to note that these three theses are independent of one another in the sense that no later thesis is entailed by an earlier one. In the first place, (2) does not entail (3), for it may be that a plurality of principles never conflict. If there are conflicts among a plurality of principles, this must be established by reference to particular cases. Moreover, even if there are conflicts, it does not follow that they can only be resolved case by case on the basis of a perception or, as for Bradley, an intuition; it is possible that they can be resolved via a priority principle. It is true that pluralists have often been skeptical about priority principles. But this skepticism is not a consequence of pluralism per se; and, as I will argue below (section 13), if there are nonarbitrary resolutions of conflict, then there is a significant sense in which priority principles cannot be avoided.

7

In the current debate, three kinds of argument have emerged as the central considerations in favor of the claim that there are genuine moral dilemmas: the argument from moral sentiment, the argument from a plurality of values, and the argument from single-value conflicts. The first of these is based on a claim about the explanation of the sentiments of regret (or remorse or guilt) that are purported to occur in the wake of practical resolutions of moral conflict. This argument first received expression in Williams's essay, "Ethical Consistency" (chapter 6), an essay that has become the focal point of much of the recent discussion.[9]

Williams's purpose is to show that, in an important respect, moral conflicts are more like conflicts of desire than conflicts of ordinary factual belief. When we act on one of two conflicting desires, he says, the rejected desire is not thereby abandoned; this can show itself in the regret we feel for what is missed. On the other hand, when we accept but one of two conflicting beliefs, the rejected belief *is* abandoned; there is no regret in losing what we now regard as false. Now, Williams argues, when two moral 'oughts' conflict and we act on one, we do not necessarily abandon the rejected 'ought'; this can show itself in the regret we feel for what was not done. Thus, in moral conflicts, like conflicts of desire but not belief, the conflicting elements cannot be eliminated. There is a moral "remainder": even if we think that we "acted for the best," the phenomenon of regret shows that it is a mistake to think that "one *ought* must be totally rejected in the sense that one becomes convinced that it did not actually apply" (p. 134).[10]

Williams anticipates the objection that such regret, if it exists at all, is either irrational or nonmoral (for instance, regret simply for doing something unpleasant). But he insists that these explanations do not cover all cases of regret: there are cases in which an agent reasonably regrets what he has done on moral grounds even though, through no fault of his own, he could not have done otherwise. The central line of response to Williams's argument, suggested here in the essays by McConnell (chapter 8), Hare (chapter 11), Conee (chapter 12), and Foot (chapter 13), is that there are other explanations of regret that do not require the postulation of genuine dilemma. Several explanations have been offered. If both courses of action would have undesirable consequences, then there might be regret for the consequences of one's action—or for being in the situation of having to perform the action—even if it was clearly the only morally right thing to do. Or, it might be that a particular person's moral code has incompatible duties, thereby generating dilemmas and regrets; however, this hardly shows that moral dilemma is an inevitable feature of all moral codes or of a rationally acceptable one. Again, the conflicting 'oughts' might be merely *prima facie* in Ross's sense; regret about failing to fulfill one of these duties could reasonably occur since, though less than actual duties, they are nonetheless in some way more than mere appearances of actual duties.[11] Finally, the conflicting 'oughts' might be *prima facie* in Hare's sense; educated to follow these principles, a person will naturally and reasonably feel regret about violating one, but that does not mean there is not a correct resolution of the conflict at the level of "critical thinking."

These objections suggest that Foot is correct in suggesting that it is a mistake "to think that the existence of feelings of regret could show anything about a remainder in cases of moral conflict" (p. 254). Such regret as may be found is relevant only if it is reasonable from a moral point of view, but this will be the case only if it is reasonable to think that there is something morally regrettable—a situation, that is, in which two genuine 'oughts' really do conflict. We can make this point and still agree with Williams about the importance of emotions for morality, for example, that regret is sometimes morally admirable. But it is another matter to suppose that the mere fact of regret in and of itself evidences, or even tends to evidence, that something *is* regrettable.

Similar considerations apply to the claim that the phenomenon of guilt (rather than regret) in the wake of the practical resolution of an apparent moral dilemma establishes that both of the conflicting 'oughts' are true; and also to the counterclaim that moral doubt in the same situation establishes that one of the conflicting 'oughts' is false.[12] That there are cases in which guilt or doubt is felt cannot be denied. But it is another question whether these sentiments are appropriately felt and why—and this question points again to the nature and validity of the values underlying the conflict.

There is, however, another way in which moral sentiments

have entered the debate on dilemma. Marcus argues that, *given* the existence of genuine moral dilemmas, guilt is an appropriate response to their practical resolution: even though, through no previous wrongdoing, it was inevitable that one would do what ought not to be done, guilt still makes sense. Moreover, Marcus argues, such guilt has an important function. It motivates us to avoid dilemmatic situations in the future, specifically, to follow the second-order principle that "one ought to act in such a way that, if one ought to do x and one ought to do y, then one can do both x and y" (p. 200). Against Marcus, Foot argues that guilt is an inappropriate response to dilemma, on the ground that in such a circumstance the agent cannot be faulted for what he did.[13] Resolution of this issue presumably would benefit from some clarification of the meaning(s) of 'guilt'. In any case, this dispute does not involve an inference from guilt to dilemma but an inference from dilemma to guilt. Hence, the objection above to Williams's argument from regret to dilemma has no parallel here.

8

A second and, to my mind, more promising line of argument for the possibility of genuine moral dilemmas does not appeal to any moral sentiments. According to this argument, there is a plurality of genuine moral values, and it is inevitable—the world being what it is, at least—that these values sometimes conflict. This argument has roots in the pluralism of both the Hegelian and the British intuitionist traditions; in the essays here it is suggested by van Fraassen (chapter 7) and especially by Lemmon (chapter 5) and Nagel (chapter 9).[14]

According to Lemmon, "there are generically different ways in which it can come to be true that we ought to do something or ought not to do something" (p. 105). Specifically, he claims, an ought statement may be based on a person's status or position, on his previous commitments, or on "moral principles." But ought statements justified in these different ways can conflict. Thus, a duty based on position may conflict with a moral principle. Situations in which a person both ought and ought not to do something, Lemmon argues, "will appear generally in the cases where these sources conflict" (p. 107). Another version of the pluralist argument is developed in more detail by Nagel. He claims that there are "conflicting and incommensurable" value claims resting on obligations (derived from relationships or undertakings), rights, utility, "perfectionist ends," and private commitments. These five values may be reduced to two categories, the personal (or "agent-centered") and the impersonal (or "outcome-centered"), but this "great division," Nagel argues, "is so basic that it renders implausible any reductive unification of ethics" (p. 179; cf., Nagel, 1972, 1980, 1986). For this reason there are genuine moral dilemmas.

There are several lines of response to this kind of argument. One is to claim that there is a single, supreme value to which all other values may be reduced and from which no conflicts may ensue. The most prominent instance of this approach is utilitarianism (see section 4 above). Pluralists find any such reductive program unrealistic: such a program cannot do justice to what they take to be the manifest diversity of human values and hence inevitably loses or distorts something essential. There is, of course, a long history of this debate in the literature of utilitarianism.

A second response is to draw a distinction between moral and nonmoral values and to argue that though there may be different and conflicting nonmoral values, there is only one moral value, and that it always takes priority over nonmoral values. Hence, though there may be genuine practical dilemmas, there are no genuine moral dilemmas. This approach is common in "Kantian" theories and is suggested here in the essays by Conee (chapter 12) and Donagan (chapter 14). As frequently developed, the idea is that morality proper contains a set of requirements and prohibitions that cannot conflict (say, along the lines of Kant's perfect duties), but that, consistent with following these rules, there is a great deal of choice (in the realm of the morally "permissible"), choice that may encounter incompatible, but nonmoral, values. From this point of view, values such as Nagel's perfectionist ends and private commitments may not be seen as "moral" at all, though this does not mean they are unimportant. At issue here, and permeating a good deal of the debate about moral dilemmas, is the broader issue of the scope of the term 'morality'. Those who reject the possibility of genuine moral dilemmas—for example, Donagan—often have a narrower conception of what considerations are morally relevant than many of those who acknowledge the possibility.

Each of these two responses involves a denial of moral pluralism. But there are also responses available from within the pluralist perspective. Thus, it may be argued that, though there is a plurality of moral values, these values do not in fact conflict. This response, however, has rarely been advanced. But it has been argued frequently that, even if there is a plurality of sometimes conflicting moral values, the conflicts nonetheless may be resolved. This may be achieved in different ways, for example (as for Ross), via a judgment about the relative stringency of the conflicting values in each particular case or (as for Rawls, 1971, p. 61) on the basis of a second-order priority principle (see section 13).[15] However, it is important to note that, when it is determined that one value overrides another, on whatever basis, two quite different kinds of resolution may be intended. On the one hand, it may mean that one of the two conflicting values does not actually apply to the case at hand, and so there is no genuine dilemma. Here 'resolution' means elimination of the conflict. On the other hand, it may mean that, though all things considered it is best to follow one value rather than another, the other value continues to apply, and so—at some level—there is a genuine dilemma. Here 'resolution' does not mean elimination of the conflict, and it is compatible with the remainder thesis urged by Williams and also with the consequence of that thesis, that regret may be an appropriate response (see section 12).

Pluralists have in any case tended to be skeptical of purported resolutions of, or methods of resolving, conflicting values. There are also two forms of skepticism that may be intended here. One is that we lack knowledge of conflict resolutions since, as is often suggested, life is simply too complicated either to make accurate judgments case by case or to devise priority principles that will cover a class of cases. However, it is consistent with this point that a resolution is, in a sense, there to be discovered, and it is simply a comment on our epistemic inadequacies that we are unable to discover it. A stronger form of skepticism, however, is suggested by Nagel's claim that some values are "incommensurable." For this implies that it is impossible in principle to resolve conflicts among them. It is not simply that we cannot discover the resolution: there is no resolution to be discovered (see section 14).

9

The third and final argument for the existence of moral dilemmas is based on the idea that a single moral value or principle can, under some conditions, conflict with itself. According to Marcus, who develops this idea (chapter 10), even if pluralism were false and there was only one moral principle, there could still be conflicts. Consider, for example, the principle of promise-keeping. A person may make two promises with the intention of keeping each and then discover that, as a result of unforeseeable and uncontrollable circumstances, it is impossible to keep both. Or consider the principle that innocent lives are to be saved. There might be an occasion in which it is possible to save one life or another but not both. In each case, a single principle enjoins incompatible actions.

There are several responses to this argument, and they parallel to a large extent the responses to the pluralist argument just considered. Thus, it may be argued, following Hare, that a principle entailing conflicting applications is but a surrogate for an ultimate moral principle, such as the principle of utility, and that, when applications conflict, authority defers to the ultimate principle, from which no conflict may be derived. Or it may be argued, as Donagan does with respect to the second example above, that morality requires only that one *or* the other life be saved, not that one *and* the other be saved. This leaves a pressing and indeed anguished practical conflict, but it is not a moral conflict.

It may also be argued, as Donagan does with respect to the first example above, that conditions implicit in the offending principle serve to eliminate the conflict. A promise is made, Donagan claims, on the understanding, first, that the promiser believes he can and (morally) may fulfill it and, second, that if it turns out he cannot or may not, then the promisee is not entitled to performance. Hence, if compliance with two promises is impossible, then either (1) the promiser could anticipate this when promising, in which case the ensuing dilemma is his fault and not the fault of morality, or (2) the promiser could not anticipate this, in which case at least one of the promisees must release him from the obligation. In neither case does the principle of promise-keeping entail a genuine moral dilemma.[16]

A final response is to argue that, when different applications of the same principle conflict, a resolution on some basis or another is nonetheless possible. But as before, 'resolution' may mean elimination of the conflict, or it may mean that, though one action is best, all things considered, the unfulfilled 'ought' persists as a moral remainder. Since a single principle is in conflict with itself, it would seem that the possibility of resolution cannot be objected to here on the basis of incommensurability (unless some additional principles were thought to come into play). However, a possible resolution may be objected to on the basis of equality (the reasons for doing A are the same as the reasons for doing B). But there is a difference between concluding, as a result of this, that A or B ought to be done, and that A ought to be done and B ought to be done. The former entails a practical dilemma, but only the latter entails a moral one.

10

The arguments and counterarguments considered so far can only be resolved through detailed articulation of moral theories along with equally detailed consideration of their application to apparent dilemmas. But there is an additional class of objections to the possibility of moral dilemma that derives from a different and purportedly more fundamental quarter, that of deontic logic. It has long been recognized that the claim that there are genuine dilemmas is inconsistent with commonly accepted principles of deontic logic. Two principles in particular have received the most attention. According to the first (agglomeration), if a person ought to do one thing and ought to do another thing, then the person ought to do both things. According to the second ("'ought' implies 'can'"), if a person ought to do something, then the person can do that thing. Now, if—as in a dilemma—S ought to do A and S ought to do B, then by agglomeration S ought to do both A and B, and from this, by "'ought' implies 'can'," it follows that S can do both A and B; however, this S cannot do if he is in a dilemma. Hence, moral dilemmas are inconsistent with the conjunction of these two principles. And there are other standardly accepted deontic principles with which dilemmas are inconsistent as well.*

To express these objections formally, let 'O' mean 'it ought to be that' and 'M' mean 'it is possible that'. The assertion that there are genuine moral dilemmas amounts to the claim that propositions of the following form can be true simultaneously:

(1) Oa;
(2) Ob;

---

*The reader may skip the formal considerations that follow and proceed to section 11 without loss of continuity.

and:

(3) $\sim M(a\ \&\ b)$.

We can express agglomeration as:

DP1 $(Op\ \&\ Oq) \supset O\ (p\ \&\ q)$,

(where 'p' and 'q' are placeholders for action-describing statements a, b, c, etc.) and "'ought' implies 'can'" as:

DP2 $Op \supset Mp$.

Now, from (1), (2), and *DP1* it follows that:

(4) $O(a\ \&\ b)$.

And from this and *DP2* it follows that:

(5) $M(a\ \&\ b)$.

But the conjunction of (5) and (3) is an explicit contradiction.

Though this argument has received the greatest amount of attention (e.g., by Williams, Marcus, Foot, and Donagan), there are also arguments (discussed by van Fraassen, McConnell, and Conee) that show the incompatibility of moral dilemma with other principles of deontic logic. According to one such principle, the "logical consequences of what ought to be, ought to be" (van Fraassen, p. 146). That is (on one formulation):

DP3 $(Op\ \&\ (p \supset q)) \supset Oq$.

Now, by standard principles of modal logic, (3) clearly entails:

(6) $\sim(a\ \&\ b)$,

and from this by the propositional calculus, we can easily arrive at:

(7) $a \supset \sim b$.

But from this, (1), and *DP3* we can conclude that:

(8) $O\sim b$.

(2) and (8) are not inconsistent, but many would find it disconcerting that something ought to be and that the negation of it ought to be as well.[17]

In any case, a fourth principle of deontic logic shows that (2) and (8) entail an inconsistency. This principle says that if something ought to be, then it is not the case that the negation of that thing ought to be. That is:

$DP4\ Op \supset\ \sim O\sim p.$

(2) and $DP4$ entail that:

(9) $\sim O\sim b$;

and this is inconsistent with (8).

Finally, it is usually thought that the fact that one thing ought to be is equivalent to the claim that it is not the case that the negation of that thing is permissible. That is:

$DP5\ Op \equiv\ \sim P\sim p,$

(where 'P' means 'it is permitted that'). From this principle and (2) it follows that:

(10) $\sim P\sim b.$

This is not inconsistent with (8), but it may seem absurd, as Conee argues (chapter 12), to suppose that something that ought to be is not permissible.

In sum: if there are genuine moral dilemmas (either of the form (1), (2) and (3), or of the form (2) and (8)), then two or more of the principles of deontic logic must be abandoned. Williams abandons agglomeration ($DP1$); Lemmon gives up "'ought' implies 'can'" ($DP2$); and similar moves are suggested by van Fraassen, Marcus, and Foot. Either move avoids the first contradiction, but neither (nor both together) avoids the second. Hence, the defender of genuine dilemmas must give up, first, either $DP1$ or $DP2$ and, second, either $DP3$ or $DP4$ (and also $DP5$ if the purported absurdity just noted is to be avoided). This might suggest to those who argue for the possibility of moral dilemmas that they ought to search for that combination of deontic principles that is easiest to abandon, while maintaining those principles that are most difficult to give up. But this, I suggest, would be a mistake. For, as I will argue presently, the plausibility of all of these principles rests on a single assumption. To question any one principle amounts to questioning this assumption and thereby the ground of all of the other principles. Hence, I am inclined to think that $DP1$ through $DP5$ stand or fall together.[18]

11

Since its origin in the work of Mally (1926) and especially in von Wright's seminal paper (1951), deontic logic has been a source of perplexity and paradox. Prominent among its quandaries are the paradoxes of Ross (1941), Prior (1954), and Chisholm (1963). In view of these and other difficulties, there can be no appeal to the principles of "standard" systems of deontic logic with anything even approaching the confidence vested in the principles of propositional and predicate logic (cf., Åqvist, 1984, pp. 608–09). It is doubtful that a deontic system with any substance can be formulated that is neutral on significant questions of moral theory (see Sayre-McCord, 1986). Accordingly, from the fact that moral dilemmas are inconsistent with commonly accepted principles of deontic logic, it cannot be concluded immediately that moral dilemmas are impossible. There must be some consideration of the rationale for these principles.

The most important factor here, at least in the historical development of deontic logic, has been a purported analogy between, respectively, the deontic modalities of obligation, prohibition, and permission and the alethic modalities of necessity, impossibility, and possibility. On this view, moral obligation is conceived as analogous to logical necessity in some important respect (and likewise for the other modalities). This analogy was explicit in von Wright (1951) and has been influential ever since. Among its significant manifestations is the idea that the laws of distribution that govern the alethic modalities have exact parallels in deontic logic (see McConnell, chapter 8). Hence, the alethic law that 'it is necessary that p and q' is equivalent to 'it is necessary that p and it is necessary that q' has its counterpart in the deontic law that 'it ought to be that p and q' is equivalent to 'it ought to be that p and it ought to be that q'. The latter, of course, entails the agglomeration principle, and in fact all the deontic principles that are inconsistent with moral dilemma (*DP1* through *DP5*) have counterparts in standard systems of alethic modal logic (e.g., system T in Hughes and Cresswell, 1972). There have been attempts, in fact, to reduce deontic logic to alethic modal logic (e.g., Anderson, 1958). The assumption that underlies these principles, then, is the claim that moral obligation is analogous to logical necessity in some important way.

Why accept this assumption? It was recognized by von Wright at the beginning that there is a significant *disanalogy* between these two modalities, namely that the alethic principles that '*p*' entails 'it is possible that *p*' and that 'it is necessary that *p*' entails '*p*' do not have plausible deontic counterparts (see Hughes and Cresswell, 1972, pp. 301–02). Moreover, von Wright has expressed further doubts of late (1983), not pertaining to moral dilemma, about the purported analogies between the alethic and deontic modalities. Hence, the analogy is not perfect in any case. Nonetheless, it does have some plausibility, and I believe the basis for this is found in the thought that moral prescriptions lay down necessary requirements for action. Even as a necessary proposition *must* be true no matter what, so it is thought that a moral prescription *must* be obeyed no matter what. If we think of moral prescriptions in this way, it seems reasonable to accept the deontic principles that conflict with dilemma. For, to consider only the first two, if it morally must be that *p* and it morally must be that *q*, then surely it morally must be that *p* and *q*; and also, if it morally must be that *p*, then surely it is possible that *p*. It would be unreasonable, on this view, for morality to make demands upon us were these principles not true (see Donagan, chapter 14).

Moral necessity, as this concept may be called, should be distinguished from both universalizability, which has to do with the scope of moral prescriptions, and objectivity, which has to do with the justification of moral prescriptions. It is not incoherent to declare an action necessary without declaring it necessary for everyone and without declaring it justified for everyone. However, moral necessity is related to the claim that moral prescriptions override all other prescriptions (see Hare, chapter 11). If moral prescriptions make certain actions necessary, then any prescription with which they conflict must yield.

The concept of moral necessity appears to be what Kant has in mind when he claims that "obligation includes *necessitation*" (p. 38). In view of this, it is not surprising both that Kant explicitly endorses "'ought' implies 'can'" (1971, p. 37) and that he declares a conflict of obligations to be impossible. It will be recalled that his argument against dilemma is based on the idea that some actions are morally necessary (section 3). The argument from the deontic principles is, in effect, an explication of the Kantian position.

12

The minimal conclusion that should be drawn from the foregoing discussion is that, if there are situations in which S ought to do A and ought to do B, but cannot do both, then 'ought' in these judgments cannot express moral necessity. For if it did, then it would be reasonable to accept the deontic principles *DP1* through *DP5*, and an inconsistency would result. If it is also supposed that a moral 'ought' can *only* express necessity, then it follows further that moral dilemmas are impossible; this is the conclusion that is often reached. However, if it is supposed that there are moral dilemmas, it is possible to circumvent this conclusion, *either* (1) by denying that moral prescriptions ever express necessity *or* (2) by claiming that there are two kinds of moral prescriptions—those that can conflict, but do not express necessity, and those that express necessity, but do not conflict. The first option appears problematic, since we do think that on some occasions a moral prescription must be obeyed no matter what. But the second option suggests a way in which the possibility of moral dilemmas might be defended.

Those who have claimed there are dilemmas have divided them into those that are resolvable (but with a "remainder") and those that cannot be resolved (because of "incommensurability"). The latter will be considered in section 14. With regard to the former, it is supposed that, even though S ought to do A and S ought to do B, there is nonetheless some sense in which, in the final analysis, (say) S ought to do B. But it is clear that the second expression of 'S ought to do B' cannot, as Williams says, "represent a mere *iteration* of" the first (p. 136). It must be made from a perspective that goes beyond the perspective of the first, though it may include that perspective. In order to take this position, it is essential to distinguish two senses of 'ought', the first used to express the horns of the dilemma, and the second used to express what, in the final analysis, ought to be done. For Williams, the second 'ought' is based on *nonmoral* as well as moral considerations and so is not, strictly speaking, a moral 'ought' at all: it is what he calls a "deliberative *ought*"(p. 135). It appears to follow from this both that there is no resolution of a moral dilemma from a purely moral perspective and that there is only one moral sense of 'ought', one that does not express necessity (see Williams, 1981a). On this interpretation, then, Williams accepts option (1) above.

But a different approach is suggested by Foot (chapter 13). She distinguishes two senses of 'ought', both of them moral senses. The first (type 1) expresses a moral 'ought' that can conflict with other moral 'oughts', thereby resulting in a dilemma. The second (type 2) expresses "the thing that is *best* morally speaking" (p. 256). For Foot, it cannot be the case that a person both ought and ought not to do something in this second sense of 'ought'; that would imply the absurdity that both courses are best. But there is no reason to deny that the first type of 'ought' can conflict with the second type. What is important here is that Foot's type 2 statements may be interpreted according to the standard deontic principles. Foot's position, then, exemplifies option (2) above. There are two kinds of moral prescriptions: those of the first kind may conflict but are not governed by the deontic principles, while those of the second kind may not conflict but are governed by these principles (cf., Chisholm, 1979; Davidson, 1970; Harman, 1975a; Searle, 1978).

In my view, and I am going beyond Foot here, the most natural terms in English to express this kind of distinction are 'ought' and 'must', taking both in a specifically moral sense.[19] Thus, I will suppose that ought-prescriptions may conflict without inconsistency, but that must-prescriptions may not conflict; and I will suppose that the deontic principles govern only the latter. A must-prescription declares what is morally best and hence what, from the moral point of view, must be done. It may be less clear, however, what is meant by an ought-prescription. If conflicts among ought-prescriptions are genuine moral conflicts, they cannot be mere appearances of must-prescriptions. An account must be given, then, of the sense of 'ought' in which it is possible that S morally ought to do A even though S morally must do B (where doing A and doing B are incompatible). One obvious approach is suggested by moral pluralism (see section 8). On this view, an ought-prescription declares, from the perspective of one among many moral values, that an action ought to be done. Hence, 'S ought to do A' is always an abbreviation of 'from the perspective of such-and-such value, S ought to do A'. (A further qualification may be employed to account for single value conflicts [see section 9].) Understood in this way, it is possible to see how ought-prescriptions may conflict without being inconsistent, while at the same time being genuine, though not conclusive, action-guiding statements. This account also suggests a way of understanding Williams's "remainder" thesis. When S does B because he must, even though he also ought to do A, when doing A

and doing *B* are incompatible, there is a clear sense in which an 'ought' persists as a moral remainder. For it remains true that *S* ought to do *A*, and this may reasonably result in regret for not having done *A* and also, perhaps, in the need to compensate.

13

In addition to the objections to moral dilemmas originating in deontic logic, there is a related set of epistemological issues raised by the possibility of dilemmas. These concern the ways in which moral dilemmas affect the cognitive status of moral judgment. Consider first those cases, as above, in which a dilemma is said to be resolvable: even though *S* ought to do *A* and ought to do *B*, when both cannot be done, nonetheless (let us say) *S* must do *B*. Given that the ought-prescriptions conflict, how is this must-prescription to be determined? Two approaches are prominent in recent literature.[20] One is that we must rely on an intuition of what, in the particular case, must be done. The other is that we must rely on a priority principle to the effect that the reason for doing *B* takes precedence over the reason for doing *A*.

It is often suggested, following Bradley and Ross, that a workable set of priority principles cannot realistically be formulated, and that, as a consequence, we can only determine what to do case by case on the basis of an immediate "intuition" or "perception" or "judgment" (see Nagel, chapter 9). The absence of a decision-procedure, Urmson writes, "leaves us with the need for an intuitive weighing up of the reasons" (1974–75, p. 119). But to put matters in this way involves a confusion between two issues: first, whether we can formulate a principle to cover a class of cases or whether we must judge each case individually; and second, whether or not our decision rests upon an intuition, be it about a class of cases or an individual case.

Consider first the claim that we must judge case by case. According to this view, in the face of a particular conflict, let us say between moral principles *P1* and *P2*, there must be a judgment such as:

(1) In *this* case, follow *P1* rather than *P2*.

However, if (1) is warranted, then there must be some feature *F*, however complex, of the case to which (1) refers in virtue of which

P1 takes priority over *P2* in that case. If there is no such feature, then (1) is arbitrary and could not be warranted, intuitively or otherwise. But if there is such a feature, then (1) involves a commitment to and may be seen as an implicit inference from:

(2) When *P1* and *P2* conflict, and *F* is present, then follow *P1*.

Of course, this is a priority principle. It need not be supposed that a priority principle is simple, declaring that in all circumstances one principle overrides the other. It may declare that *P1* overrides *P2* in some circumstances and that *P2* overrides *P1* in others. Even if *F* is a configuration of characteristics so unusual that it is extremely unlikely that there will be any other actual cases where *F* is present, (2) is still a priority principle, in that it refers to a kind of case and not simply a particular case.[21]

Thus, in cases where conflicts admit a nonarbitrary resolution, priority principles cannot be avoided. The further issue concerning our knowledge of these principles is obviously a difficult one, as are all questions concerning moral knowledge. However, it is not difficult to see that one reason typically offered for supposing that "intuition" must be involved is not compelling. Thus, it is often suggested that the moral life is so complicated that no simple set of priorities can realistically be formulated; from this it is said to follow that priority principles will have to have such a narrow scope that they cannot be formulated in advance and hence will have to be formulated case by case. But if this is so, the argument continues, there will be no alternative at the moment of decision but to base the principle on intuition. However, even if the complexity of the moral life prevents formulation of a workable decision-procedure for all future cases, all that follows is that it is sometimes necessary to act under conditions of uncertainty. There is nothing in these considerations that suggests that nonintuitive knowledge of priority principles is impossible; at most they suggest that sometimes it could be available only retrospectively.

14

Attention has been restricted thus far to dilemmas that are said to have a resolution. But it has also been claimed that there are dilemmas that have no resolution. Williams argues that there are dilemmas in which "the notion of 'acting for the best' may very well lose its content" (p. 123). According to Nagel, "there are true practical dilemmas that have no solution" (p. 180). Similar claims are made by van Fraassen and Foot. The principal argument for irresolvable dilemmas is that conflicting values are sometimes "incommensurable" and that in such situations nothing can be said from a rational viewpoint about what in the final analysis is best (though a great deal may be said to rationally establish the dilemma in the first place).

The concept of incommensurability is a familiar if troublesome concept from recent discussions in the philosophy of science. What gives it plausibility in the moral realm is our sense of bafflement in the face of questions such as whether the good of friendship is more or less worthwhile than the good of justice, or whether the good of integrity is better or worse than the good of some worthy objective. There is no obvious common ground for such comparative judgments, and this suggests that these values are incommensurable. It is doubtful in any case that we can always feel fully confident about the rationality of such judgments. At the same time, though, it is surely sometimes the case that we do feel quite confident, as when the point of justice is of great significance and the issue of friendship is minute. Yet it is problematic to suppose both that sometimes one value takes priority over another and that sometimes it does not, on account of incommensurability. If two values are really incommensurable, it is unclear how it can ever be the case that one takes priority over the other. It may be, of course, that some values are commensurable and others are not, but it is another matter to suppose that the same two values are commensurable on some occasions and not on others. On the face of it this is a puzzling idea. What are thought to be cases of incommensurable values sometimes may be cases in which values, though commensurable, do not have a sufficiently determinate and precise measure of worth attached to them to give us confidence in comparative judgments when calls are close. This is a rather different idea than incommensurability, though it may also render some dilemmas irresolvable.[22]

In any case, if there are irresolvable dilemmas, then it is not always the case that there is one action that is morally the best (in my terminology, that must be done). This puts an obvious limitation on the extent to which moral judgment can be said to be objective. However, from the fact that in a given situation it is not the case that one action is the best, it does not follow that in that situation any action is as good or as bad as any other. It may still be that some actions are better than others. In general, whenever there is a plurality of considerations relevant to a question but indeterminate in their relative importance—as, for example, in a hiring decision—we may be faced with situations in which, though there is no one right answer, some answers are clearly better than others. It has even been argued, though controversially, that scientific questions are sometimes of this nature (e.g., see Kuhn, 1977, and Putnam, 1981, pp. 147–48).

15

A final epistemological issue is raised by Williams's claim, in the essay here and elsewhere (1966), that the existence of moral dilemmas is incompatible with, and hence provides an argument against, moral realism (or cognitivism). In evaluating this claim, it is important to distinguish different senses in which the term 'realism' may be used. In one sense of the term, realism is the view that the truth-value of a statement or judgment is determined by the world, where the world is taken to be something independent of human reason, perception, will, desire, and the like. The meaning of 'realism' in this sense is intuitively obvious to many but notoriously difficult to elucidate. In any case, it has played little direct role in this debate except insofar as it has been associated with other senses of 'realism', in particular (1) that conflicting statements cannot both be true and (2) that every statement must be either true or false.

With regard to (1), Williams argues that, since conflicting ought statements can both be affirmed, as evidenced by regret, then moral realism in this sense cannot be correct. This argument is challenged by Foot. She agrees that there is a sense in which there are moral dilemmas, but she rejects the inference from this to the denial of moral realism or cognitivism. In particular, she argues that the cognitivist can "simply *allow* the truth of 'I ought to do a' and 'I ought to do $\sim a$'" (p. 262). As long as at least one of the ought statements is type 1, they can be both true and conflicting, and there can be perfectly straightforward and objective evidence for each. Type 2 ought statements, on the other hand, cannot be both true and conflicting: it cannot be the case that both a and $\sim a$ are morally the best. However, if we are referring to a conflict among 'oughts', at least one of which is type 1, then there is no threat to realism. About this, Foot is clearly correct. Realism requires that *inconsistent* statements cannot both be true. But realism in this sense does not pre-clude the possibility that conflicting statements can both be true, as long as they are consistent. In my terms, ought-prescriptions may enjoin conflicting actions yet remain consistent, since the deontic principles do not apply. Hence, conflicting ought-prescriptions pose no threat to realism. On the other hand, if must-prescriptions were to conflict, then realism would be threatened, since the deontic prin-ciples would apply and an inconsistency would result.

However, if we understand 'realism' as (2), the view that a statement must be either true or false, then the existence of *irresolv-able* moral dilemmas might be thought to be incompatible with real-ism. For if there are dilemmas that are irresolvable, because of incommensurability, then it might be supposed that there are situ-ations in which a statement about what is morally best has no truth-value. With respect to irresolvable dilemmas, Foot argues that no final judgment should be made: we should "declare that the two are incommensurable, so that we have nothing to say about the overall merits of *a* and *b*" (p. 267). We might decline judgment in recog-nition of our epistemological limits, but also, she argues, "because there is no truth of the matter." This appears to be a rejection of realism in the present sense. However, against this, it might be argued that situations in which incommensurable values conflict do not deprive moral judgments of truth-value; all that follows is that no one thing is morally best (must be done, in my terminology) and hence that every judgment to the effect that one thing is best is *false*. In such a situation, what is *true* is that there are at least two actions that are morally permissible. But there are no moral judgments that lack truth-value altogether.[23]

Notes

1. For other literary treatments of moral conflict, see James's *The Golden Bowl*, discussed by Nussbaum, 1983; Melville's *Billy Budd*, dis-cussed by Winch, 1965; and Styron's *Sophie's Choice*, discussed by Green-span, 1983.

2. See Aquinas 1964–75, I–II, 19, 6 *ad* 3; II–II, 62, 2 *obj.* 2; and III, 64, 6 *ad* 3; 1952–54, 17, 4 *ad* 8; also Donagan, chapter 14, and 1977, chapter 5.

3. See, for instance, Aune, 1979, pp. 191–97; Herman, 1985, pp. 420–22; Nell, 1975, pp. 132–37; and Nussbaum, 1985, pp. 242–44.

4. See, for instance, Aune, 1979, pp. 188–94; Donagan, 1977, pp. 154–57; Gregor, 1963, pp. 95–112; and Nell, 1975.

5. For a detailed attempt to deal with apparent moral conflicts from a viewpoint with some kinship to Kant, see Gewirth, 1978, pp. 338–54. Gewirth's approach to moral conflict is discussed in Bambrough, 1984; Raphael, 1984; and Singer, 1984; Gewirth responds in 1984, especially pp. 249–53.

6. See also Sidgwick, 1981, p. 422, and, more recently, Hare, 1981, part 2; Nowell-Smith, 1972–73, p. 417; Sartorius, 1975, pp. 10–11; and Williams, 1972, pp. 92–93.

7. Moral conflict and utilitarianism are discussed in Bronaugh, 1975; Griffin, 1982; Hoag, 1983; Lyons, 1978; McConnell, 1981b; Rabinowicz, 1978; Slote, 1985; Stubbs, 1981; and Williams, 1973b.

8. Also of interest in this respect are Dewey and Tufts, 1932, part 2, and Hartmann, 1932, volume 2.

9. In addition to the papers included in this volume, see Atkinson, 1965; Guttenplan, 1979–80; Harrison, 1979; and Trigg, 1971. Also, especially important by Williams in this connection are 1966, 1976a, 1978, and 1979.

10. Cf., De Sousa, 1974; Jackson, 1985; Phillips and Price, 1967; Phillips and Mounce, 1970, chapter 8; and Phillips, 1982, chapter 3.

11. In fact, Ross himself suggests that feeling "compunction" may be appropriate in this situation (p. 93).

12. The relation of moral dilemma and guilt is discussed by van Fraassen, chapter 7; Marcus, chapter 10, and Foot, chapter 13; the argument based on moral doubt is discussed by McConnell, chapter 8.

13. See also Anderson, 1985, Conee, chapter 12; and Greenspan, 1983.

14. Moral pluralism, in a variety of forms, has been widely proposed in recent years. See in particular Berlin, 1969, 1978; Blum, 1980; Hampshire, 1977, 1983; Hook, 1960; MacIntyre, 1981; Mallock, 1967; McCloskey, 1969; Nozick, 1981, chapter 5, section 1; Nussbaum, 1985, 1986; Platts, 1979, chapter 10; Rawls, 1971; Sen, 1985; Taylor, 1982; Urmson, 1974–75; Wiggins, 1976; and Williams, 1979.

15. The addition of exception clauses within otherwise conflicting first-order principles may be equivalent to the second approach (see Hare, chapter 11), but it is not a natural way to express the second sense of 'resolution' described in this paragraph.

16. Donagan's full response is somewhat more involved than this. See chapter 14.

17. Sometimes the claim that there are moral dilemmas is expressed not by saying that there are true propositions of the form (1), (2), and (3), but by saying that there are true propositions of the form (2) and (8). If we assume *DP3,* the original formulation entails the latter. Also, if we assume *DP1* and *DP2,* then (2) and (8) entail the absurdity that the conjunction of b and $\sim b$ is possible.

18. Versions of *DP1* through *DP5* are discussed in Føllesdal and Hilpinen (1971) on pp. 3 and 11, 20, 4, 13, and 8, respectively. The most controversial principle, independent of the debate on dilemma, is "'ought' implies 'can'" (*DP2*). It is accepted by von Wright (1971, p. 163) but not by Hintikka (1971, pp. 83–84). For discussion of this principle in relation to moral dilemma, see Margolis, 1967; Rescher, forthcoming; Sinnott-Armstrong, 1984; Trigg, 1971; and Zimmerman, forthcoming. For additional discussion of deontic logic and dilemma, see Castañeda, 1966; Lemmon, 1965; McConnell, 1976; Sinnott-Armstrong, forthcoming; and Swank, 1985.

19. The distinction between 'ought' and 'must' is discussed in Cavell, 1979, part 3; Lemmon, chapter 5; Wertheimer, 1972; White, 1975; and Williams, 1982.

20. In addition to the discussions in sections 5, 6, 8, and 9, see McDowell, 1979; Nozick, 1981, chapter 5, part 3; and Wiggins, 1975–76.

21. This argument does not depend, as Steiner (1983, p. 63) seems to suppose, on moral judgments being universalizable, meaning that they apply to anyone in the same situation; it depends only on (1) being based on some feature of the case in question. For a different view see Dancy (1983, p. 545–46). The relationship between universalizability and moral conflict is discussed in Anderson, 1985; Kolenda, 1975; MacIntyre, 1957; MacLean, 1984; Marcus, chapter 10; and Winch, 1965.

22. Cf., Griffin, 1977; Sinnott-Armstrong, 1985; Wiggins, 1976; and Williams, 1979.

23. In addition to the articles included in this volume, moral dilemma and moral realism are considered in Guttenplan, 1979–80; Harrison, 1979; Platts, 1979, chapter 10; Sinnott-Armstrong, unpublished, and Tännsjö, 1985.

1

Moral Duties

Immanuel Kant

Introduction to *The Metaphysic of Morals*

III. On the Division of a Metaphysic of Morals[1]

Every legislation contains two elements (whether it prescribes external or inner actions, and whether it prescribes these a priori by mere
reason or by the choice of another person): *first* a law, which sets
forth as objectively necessary the action that ought to take place, i.e.
which makes the action a duty; *secondly* a motive, which joins a
ground determining choice to this action *subjectively* with the
thought of the law. Hence the second element consists in this: that
the law makes the duty into the motive. The first presents an action
as a duty, and this is a merely theoretical recognition of a possible
determination of the power of choice, i.e. of a practical rule. The

From *The Doctrine of Virtue: Part II of the Metaphysic of Morals,* trans. Mary J.
Gregor (Philadelphia: University of Pennsylvania Press, 1971), pp. 16–25 and 43–
56. Reprinted with the permission of the translator Mary J. Gregor. The title is provided by C.W.G. Pagination in square brackets refers to that of the Prussian Academy edition.

second connects the obligation to act in this way with a ground for determining the subject's power of choice as such.

Legislation can therefore be identified by the motives it uses (even if, with regard to the action that it makes a duty, it coincides with another legislation, [218] as when both prescribe external actions). The legislation that makes an action a duty and also makes duty the motive is *ethical.* But the legislation that does not include the motive in the law and so permits a motive other than the Idea of duty itself is *juridical.* It is clear that in the case of juridical legislation this motive which is something other than the Idea of duty must be drawn from the pathological determining grounds of choice, the inclinations and aversions, and, among these, from aversion, since the nature of legislation is to necessitate rather than to invite.

The mere conformity or nonconformity of an action with the law, without reference to the motive of the action, is called its *legality* (lawfulness). But that conformity in which the Idea of duty contained in the law is also the motive of the action is called its *morality.*

Duties in accordance with juridical legislation can be only external duties, since this legislation does not require that the inner Idea of the duty be of itself the ground determining the agent's choice; and since it still needs a motive appropriate to the law, it can connect only external motives with the law. But ethical legislation, while it makes inner actions duties as well, does not exclude external actions: it is concerned with all duties in so far as they are duties. But just because ethical legislation includes in its law the inner motive of the action (the Idea of duty), which must not be considered in outer legislation, it cannot be outer *legislation* (not even that of a divine will). It does, however, admit into itself duties which are based on another (outer) legislation, by making them, *as duties,* motives in ethical legislation.

From this it can be seen that all duties, merely because they are duties, belong to ethics. But it does not follow that their legislation is always contained in ethics: in the case of many duties it lies outside ethics. Thus ethics commands that I fulfill a contract I have entered into, even if the other party could not compel me to do it. But it accepts the law *(pacta sunt servanda)* and the duty corresponding to it from the doctrine of Law, as [219] already given there. Accordingly, the legislation that promises agreed to must be kept does not lie in ethics but rather in *ius.* All that ethics teaches about it is that if the motive which juridical legislation connects with that duty—external compulsion—is let go, the Idea of duty by itself must be the sufficient motive. For if this were not the case—were the legislation itself not juridical and the duty springing from it not really a juridical duty (as distinguished from a duty of virtue)—then faithful performance (of the promises made in a contract) would be put in the same class with acts of benevolence and the obligation to them. And this must not happen. To keep one's promises is no duty of virtue but a juridical duty, one which we can be compelled to fulfill. But it is still a virtuous action (a proof of virtue) to keep our promises even when we need not *worry* about compulsion. Hence what distinguishes the doctrine of Law from the doctrine of virtue is not so much their different duties as rather the different kinds of legislation which connect one or the other motive with the law.

Ethical legislation *cannot* be external (even if all the duties in question are external ones): juridical legislation, like the duties, can be external. Thus it is an external duty to keep the promises made in a contract; but the command to do this merely because it is a duty and without regard for any other motive belongs only to *inner* legislation. So the reason for assigning an obligation to ethics is not that the duty is of a particular kind (a particular kind of action to which we are obligated)—for there are external duties in ethics as well as in Law—but rather that the legislation is inner legislation and there can be no external lawgiver. Thus duties of benevolence, although they are external duties (obligations to external actions), belong to ethics, because only inner legislation can enjoin them.— Ethics does have its special duties as well (for example, duties to oneself), but it also has duties in common with Law. What ethics does not have in common with Law is only the kind of *obligation* [to these duties]. For the characteristic property of ethical legislation is that it commands us to perform actions merely because they are duties and to make the principle of duty itself the sufficient motive of our choice. [220] Thus there are, indeed, many *directly ethical* duties, but inner legislation makes all other duties indirectly ethical.

IV. Concepts Preliminary to the Metaphysic of Morals (Philosophia Practica Universalis)

The concept of *freedom* is a pure rational concept, which is therefore transcendent for theoretical philosophy: because the concept of freedom is such that no example adequate to it can ever be given in any possible experience, freedom is not an object of our possible theoretical knowledge. For speculative reason the concept of freedom can have no validity as a constitutive principle but solely as a regulative and, indeed, merely negative principle. But in reason's practical use the concept of freedom proves its reality through practical principles which, as laws of a causality of pure reason which is independent of all empirical conditions (of sensibility as such), determine choice and prove the existence in us of a pure will in which moral concepts and laws have their source.

On practical reason's positive concept of freedom there are based unconditioned practical laws called *moral* laws. But since our power of choice is sensuously affected and so does not of itself conform with the pure will but often opposes it, in relation to us these moral laws are *imperatives* (commands or prohibitions) and, indeed, categorical (unconditioned) imperatives. As such they are distinguished from technical imperatives (precepts of art [*Kunstvorschriften*]), which are always merely conditioned commands. According to categorical imperatives certain actions are *permissible* or *impermissible*, i.e. morally possible or impossible, while some of these actions or their contraries are morally necessary, i.e. obligatory. From these arises the concept of a duty whose fulfillment or transgression is connected with pleasure or pain of a special kind (of moral *feeling*). But in discussing practical laws of reason we do not take this feeling into account, since it does not concern the *ground* of these laws but only the subjective *effect* which they have on our mind when they determine our power of choice. And this effect can differ from one subject to the next without objectively (that is, in the judgment of reason) adding to or detracting from the validity or influence of these laws. [221]

The following concepts are common to both parts of the Metaphysic of Morals.

Obligation is the necessity of a free action under a categorical imperative.

An imperative is a practical rule which *makes* necessary an action that is in itself contingent. An imperative differs from a practical law in that a law, when it states that an action is necessary, takes no account of whether the action is also *subjectively* necessary on the agent's part (he could be a holy being) or whether the action is in itself contingent (as in the case of man). If the action is subjectively necessary there is no imperative. Hence an imperative is a rule which makes a subjectively contingent action necessary and thus shows that the agent is one who must be *necessitated* to conform with this rule.—The categorical (unconditioned) imperative views the action as objectively necessary and necessitates the agent to it immediately, by the mere thought of the action itself (i.e. of its form), and not mediately, by the thought of an *end* to be attained by the action. The only practical doctrine which can bring forth instances of such imperatives is the one that lays down obligation (the doctrine of morality). All other imperatives are *technical* imperatives; they are, one and all, conditioned. The *ground* of the possibility of categorical imperatives is this: that they are based simply on the *freedom* of the power of choice, not on any other characteristic of choice (by which it can be subjected to a purpose).

An action which is not contrary to obligation is *permissible (licitum);* and this freedom, which is not limited by any opposing imperative, is called a moral title *(facultas moralis).* From this the definition of the *impermissible (illicitum)* is self-evident.

A *duty* is an action to which we are obligated. It is, accordingly, the matter of obligation, and we can be obligated in different ways to one and the same duty (that is, to one and the same action which is a duty).

In so far as the categorical imperative asserts an obligation with respect to certain actions, it is a morally-practical [222] *law.* But since obligation includes *necessitation* as well as the practical necessity that a law as such expresses, a categorical imperative is a law which either commands or prohibits; it sets forth as a duty the commission or omission of an action. An action that is neither commanded nor forbidden is merely *permissible,* since there is no law to limit one's freedom (moral title) to perform it, and so no duty with regard to the action. An action of this kind is called morally indifferent *(indifferens, adiaphoron, res merae facultatis).* The question can be raised *whether there are* morally indifferent actions and, if so, whether we must admit permissive law *(lex permissiva),* in addition to commands and prohibitions *(lex praeceptiva, lex mandati* and *lex prohibitiva, lex vetiti),* in order to account for this moral title to act or refrain from acting as one pleases. In that case, the action to which the moral title refers would not be indifferent *(adiaphoron)* in every case; for according to moral laws, no special law is needed for an action that is always indifferent.

An action is called a *deed* in so far as it comes under obligatory laws and hence in so far as it is referred to the freedom of the agent's power of choice. The agent is considered the *author* of the effects of his deed, and these, along with the action itself, can be *imputed* to him if, before he acts, he knows the law by virtue of which they come under an obligation.

A *person* is a subject whose actions can be *imputed* to him. *Moral* personality is thus the freedom of a rational being under moral laws. (Psychological personality is merely the power to become conscious of one's self-identity at different times and under the different conditions of one's existence.) From this it follows that a person is subject to no other laws than those which he (either alone or at least along with others) gives himself.

A *thing* is something that is not susceptible of imputation. Thus any object of free choice which itself lacks freedom is called a thing *(res corporalis)*.

A deed is *right* or *wrong (rectum aut minus rectum)* when it conforms with duty or is contrary to duty (*factum licitum aut* [223] *illicitum*); the duty itself, so far as its content or its source is concerned, may be of any kind whatsoever. A deed contrary to duty is called a *transgression (reatus)*.

An *unintentional* transgression which can still be imputed to the agent is called a mere *fault (culpa)*. An *intentional* transgression (i.e. one which the agent knows is a transgression) is called a *crime (dolus)*. What is right in accordance with external laws is called *just (iustum)*: what is wrong, unjust *(iniustum)*.

A *conflict of duties (collisio officiorum s. obligationum)* would be a relation of duties in which one of them would annul the other (wholly or in part).—But a *conflict of duties* and obligations is inconceivable *(obligationes non colliduntur)*. For the concepts of duty and obligation as such express the objective practical *necessity* of certain actions, and two conflicting rules cannot both be necessary at the same time: if it is our duty to act according to one of these rules, then to act according to the opposite one is not our duty and is even contrary to duty. But there can, it is true, be two *grounds* of obligation *(rationes obligandi)* both present in one agent and in the rule he lays down for himself. In this case one or the other of these grounds is not sufficient to oblige him *(rationes obligandi non obligantes)* and is therefore not a duty.—When two such grounds conflict with each other, practical philosophy says, not that the stronger obligation takes precedence *(fortior obligatio vincit),* but that the stronger *ground of obligation* prevails *(fortior obligandi ratio vincit).*

Obligatory laws that can be given in outer legislation are called *external* laws *(leges externae)* in general. If their power to obligate can be recognized a priori by reason, even apart from outer legislation, they are *natural* external laws. But if actual outer legislation is needed to make them obligatory (and so to make them laws), they are called *positive* laws. We can therefore conceive an outer legislation which would contain only positive laws; but this would still presuppose a natural law establishing the authority of the legislator (i.e. his moral title to obligate others by his mere act of choice). [224]

The principle that makes certain actions duties is a practical law. The rule that the agent himself makes his principle on subjective grounds is called his *maxim.* Thus different men can have quite different maxims with regard to the same law.

The categorical imperative, which as such only expresses what obligation is, reads: act according to a maxim which can, at the same time, be valid as a universal law.—You must, therefore, begin by looking at the subjective principle of your action. But to know whether this principle is also objectively valid, your reason must subject it to the test of conceiving yourself as giving universal law through this principle. If your maxim qualifies for a giving of universal law, then it is objectively valid.

The simplicity of this law, compared with its great and manifold implications, must seem astonishing at first. So must its imperious authority, in view of the fact that the law, despite its authority, carries no perceptible motive with it. But while we are wondering at the power of our reason to determine choice by the mere Idea that our maxim qualifies for the *universality* of a practical law, we realize that these same practical (moral) laws first make known and establish beyond all doubt a property of our power of choice—its freedom—which speculative reason would never have hit upon by itself, either from a priori grounds or by any experience whatsoever, and which, when it is attained, can never be proved possible on theoretical grounds. Then we are less astonished to find that these laws, like mathematical postulates, are *indemonstrable* and yet *apodictic,* and at the same time to see a whole field of practical knowledge open before us, where theoretical reason, with this same Idea of freedom—indeed with all its other Ideas of the supersensible—must find everything closed tight against it.—The conformity of an action with the law of duty is its *legality (legalitas):* the conformity of the maxim of the action with the law is the *morality (moralitas)* of the action. A *maxim* is the *subjective* principle of action, the principle which the subject himself makes his rule (how he chooses to act). The principle of duty, on the other hand, is the principle that reason prescribes to him absolutely and so objectively (how he *ought* to act). [225]

The first principle of morality is, therefore: act according to a maxim which can, at the same time, be valid as universal law.— Any maxim which does not so qualify is contrary to morality.

．　．　．　．　．　．　．

Introduction to "The Doctrine of Virtue"

III. On the Ground for Conceiving an End Which Is at the Same Time a Duty

An *end* is an *object* of free choice, the thought of which determines the power of choice to an action by which the object is produced. [384] Every action, therefore, has its end; and since no one can have an end without *himself* making the object of choice into an end, it follows that the adoption of any end of action whatsoever is an act of *freedom* on the agent's part, not an operation of *nature*. But if this act which determines an end is a practical principle that prescribes the end itself (and therefore commands unconditionally), not the means (and so not conditionally), it is a categorical imperative of pure practical reason.[2] It is, therefore, an imperative which connects a *concept of duty* with that of an end as such.

Now there must be such an end and a categorical imperative corresponding to it. For since there are free actions there must also be ends to which, as their object, these actions are directed. But among these ends there must also be some that are at the same time (that is, by their concept) duties.—For were there no such ends, then all ends would be valid for practical reason only as means to other ends; and since there can be no action without an end, a *categorical* imperative would be impossible. And this would do away with all moral philosophy.

Thus we are not speaking here of the ends man sets for himself according to the sensuous impulses of his nature, but of the objects of free choice under its laws—objects man *ought to adopt* as ends.

The study of the former type of ends can be called the technical (subjective) doctrine of ends: it is really the pragmatic doctrine of ends, comprising the rules of prudence in the choice of one's ends. The study of the latter type of ends, however, must be called the moral (objective) doctrine of ends. But this distinction is superfluous here, since the very concept of moral philosophy already distinguishes it clearly from the doctrine of nature (in this case anthropology) by the fact that anthropology is based on empirical principles, while the moral doctrine of ends, which treats of duties, is based on principles given a priori in pure practical reason.

IV. What Ends Are Also Duties?

They are *one's own perfection* and the *happiness of others.*

We cannot interchange perfection and happiness here. In other words, *one's own happiness* and the *perfection of other men* cannot be made into obligatory ends of the same person. [385]

Since every man (by virtue of his *natural* impulses) has *his own happiness* as his end, it would be contradictory to consider this an obligatory end. What we will inevitably and spontaneously does not come under the concept of *duty,* which is *necessitation* to an end we adopt reluctantly. Hence it is contradictory to say that we are *under obligation* to promote our own happiness to the best of our ability.

In the same way, it is contradictory to say that I make another person's *perfection* my end and consider myself obligated to promote this. For the *perfection* of another man, as a person, consists precisely in *his own* power to adopt his end in accordance with his own concept of duty; and it is self-contradictory to demand that I do (make it my duty to do) what only the other person himself can do.

V. Clarification of These Two Concepts

A. ONE'S OWN PERFECTION

The word *"perfection"* is open to many misinterpretations. Perfection is sometimes understood as a concept belonging to transcendental philosophy—the concept of the *totality* of the manifold which, taken together, constitutes a thing. Then again, in so far as it belongs to *teleology* it is taken to mean the adequacy of a thing's qualities to an *end*. Perfection in the first sense could be called *quantitative* (material) perfection: in the second, *qualitative* (formal) perfection. The quantitative perfection of a thing can be only one (for the totality of what belongs to a thing is one). But one thing can have a number of qualitative perfections, and it is really qualitative perfection that we are discussing here.

When we say that man has a duty to take as his end the perfection characteristic of man as such (of humanity, really), we must locate perfection in what man can bring into being by his actions, not in the mere gifts he receives from nature; for otherwise it would not be a duty to make perfection an end. This duty must [386] therefore be the *cultivation* of one's *powers* (or natural capacities), the highest of which is *understanding,* the power of concepts and so too of those concepts that belong to duty. At the same time this duty includes the cultivation of one's *will* (moral attitude) to fulfill every duty as such. 1) Man has a duty of striving to raise himself from the crude state of his nature, from his animality *(quoad actum³)* and to realize ever more fully in himself the humanity by which he alone is capable of setting ends: it is his duty to diminish his ignorance by education and to correct his errors. And it is not merely technically-practical reason that *counsels* him to acquire skill as a means to his further aims (of art [*Kunst*]): morally-practical reason *commands* it absolutely and makes this end his duty, that he may be worthy of the humanity in him. 2) Man has a duty of cultivating his *will* to the purest attitude of virtue, in which the law is the motive as well as the norm for his actions and he obeys it from duty. This is the duty of striving for inner morally-practical perfection. Since this perfection is a feeling of the influence which the legislative will within man exercises on his power of acting in accordance with this will, it is called *moral feeling*—a special *sense (sensus moralis),* as it were. It is true that moral sense is often misused in a visionary way, as if (like Socrates' genius) it could precede or even dispense with reason's judgment. Yet it is a moral perfection, by which one makes each particular obligatory end one's object.

B. THE HAPPINESS OF OTHERS

By a tendency of his nature man inevitably wants and seeks his own happiness, i.e. contentment with his state along with the assurance that it will last; and for this reason one's own happiness is not an obligatory end.—Some people, however, invent a distinction between moral happiness, which they define as contentment with our own person and moral conduct and so with what we *do,* and natural happiness, which is satisfaction with what nature bestows and so with what we *enjoy* as a gift from without. (I refrain here from censuring a misuse of the word "happiness" which already involves a contradiction.) It must therefore be noted that the feeling of moral happiness belongs only under the [387] preceding heading of perfection; for the man who is said to be happy in the mere consciousness of his integrity already possesses the perfection defined there as the end which it is also his duty to have.

When it comes to my pursuit of happiness as an obligatory end, this must therefore be the happiness of *other* men, *whose* (permissible) *ends I thus make my own ends as well.* It is for them to decide what things they consider elements in their happiness; but I am entitled to refuse some of these things if I disagree with their judgments, so long as the other has no right to demand a thing from me as his due. But time and again an alleged *obligation* to attend to *my own* (natural) happiness is set up in competition with this end, and my natural and merely subjective end is thus made a duty (an obligatory end). Since this is used as a specious objection to the division of duties made above (in IV), it needs to be set right.

Adversity, pain, and want are great temptations to transgress one's duty. So it might seem that prosperity, strength, health, and well-being in general, which check the influence of these, could also be considered obligatory ends which make up the duty of promoting *one's own* happiness, and not merely the happiness of others.—But then the end is not the agent's happiness but his morality, and happiness is merely a means for removing obstacles to his morality—a *permissible* means, since no one has a right to demand that I sacrifice my own ends if these are not immoral. To seek prosperity for its own sake is no direct duty, but it can well be an indirect duty: the duty of warding off poverty as a great temptation to vice. But then it is not my happiness but the preservation of my moral integrity that is my end and also my duty.

VI. Ethics Does not Give Laws for Actions (Ius Does That), But Only for the Maxims of Actions

The concept of duty stands in immediate relation to a *law* (even if we abstract from all ends, as its matter). [388] We have already indicated how, in that case, the formal principle of duty is contained in the categorical imperative: "So act that the maxim of your action could become a universal law." Ethics adds only that this principle is to be conceived as the law of *your own will* and not of will in general, which could also be the will of another. In the latter case the law would prescribe a juridical duty, which lies outside the sphere of ethics.—Maxims are here regarded as subjective principles which merely *qualify* for giving universal law, and the requirement that they so qualify is only a negative principle: not to come into conflict with a law as such.—How then can there be, beyond this principle, a law for the maxims of actions?

Only the concept of an obligatory *end,* a concept that belongs exclusively to ethics, establishes a law for the maxims of actions by subordinating the subjective end (which everyone has) to the objective end (which everyone ought to adopt as his own). The imperative: "You ought to make this or that (e.g. the happiness of another) your end" is concerned with the matter of choice (an object). Now no free action is possible unless the agent also intends an end (which is the matter of choice). Hence, when there is an obligatory end, the maxim of the action, in so far as the action is a means to the end, need only qualify for a possible giving of universal law. As opposed to this, it is the obligatory end that can make it a law to have such a maxim, since for the maxim itself the mere possibility of harmonizing with a giving of universal law is already sufficient.

For the maxims of actions can be *arbitrary,* and come only under the formal principle of action, the limiting condition that they qualify for giving universal law. A *law,* however, does away with this arbitrary element in actions, and by this it is distinguished from *counsels* (which seek to know merely the most appropriate means to an end). [389]

VII. Ethical Duties Are of Wide Obligation, Whereas Juridical Duties Are of Narrow Obligation

This proposition follows from the preceding one; for if the law can prescribe only the maxim of actions, not actions themselves, this indicates that it leaves a play-room *(latitudo)* for free choice in following (observing) the law, i.e. that the law cannot specify precisely what and how much one's actions should do toward the obligatory end.—But a wide duty is not to be taken as a permission to make exceptions to the maxim of actions, but only as a permission to limit one maxim of duty by another (e.g. love of one's neighbour in general by love of one's parents)—a permission that actually widens the field for the practice of virtue.—As the duty is wider, so man's obligation to action is more imperfect; but the closer to *narrow* duty (Law) he brings the maxim of observing this duty (in his attitude of will), so much the more perfect is his virtuous action.

Imperfect duties, accordingly, are only *duties of virtue*. To fulfill them is *merit* (*meritum* = +a); but to transgress them is not so much *guilt* (*demeritum* = −a) as rather mere *lack of* moral *worth* (= 0), unless the agent makes it his principle not to submit to these duties. The strength of one's resolution, in the first case, is properly called only *virtue (virtus)*; one's weakness, in the second case, is not so much *vice (vitium)* as rather mere *want of virtue*, lack of moral strength *(defectus moralis)*. (As the word *Tugend* [virtue] comes from *taugen* [to be fit for], so *Untugend* [lack of virtue] comes from *zu nichts taugen* [to be worthless].) Every action contrary to duty is called a *transgression (peccatum)*. It is when an intentional transgression has been adopted as a basic principle that it is properly called *vice (vitium)*.

Although the conformity of actions with Law (being a law-abiding man) is not meritorious, the conformity with Law of the maxim of such actions regarded as duties, i.e. *reverence* for Law, is *meritorious*. For by this we make the right of humanity, or also the rights of men, our *end* and widen our concept of duty beyond the notion of *what is due (officium debiti)*, since [390] another can demand by right that my actions conform with the law, but not that the law be also the motive for my actions. The same holds true of the universal ethical command: do your duty from the motive of duty. To establish and quicken this attitude in oneself is, again, *meritorious;* for it goes beyond the law of duty for actions and makes the law in itself the motive also.

Hence these duties, too, are of wide obligation. With regard to wide obligation there is present a subjective principle that brings its ethical *reward*—a principle which, to assimilate the concepts of wide and narrow obligation, we might call the principle of receptiveness to this reward in accordance with the law of virtue. The reward in question is a moral pleasure which is more than mere contentment with oneself (this can be merely negative) and which is celebrated in the saying that by this consciousness virtue is its own reward.

If this merit is a man's merit in relation to other men for promoting their natural and so universally recognized end (for making their happiness his own), it could be called *sweet* merit; for consciousness of it produces a moral gratification in which men are prone to *revel* by sympathetic feeling. But *bitter* merit, which comes from promoting the true welfare of others even when they fail to recognize it as such (when they are unthankful and unappreciative), usually has no such reaction. All that it produces is *contentment* with oneself. But in this case the merit would be still greater.

VIII. Exposition of Duties of Virtue as Wide Duties

1. ONE'S OWN PERFECTION AS AN OBLIGATORY END

a) *Natural* perfection is the *cultivation* of all one's *powers* for promoting the ends that reason puts forward. That natural perfection is a duty and so an end in itself, and that the cultivation of our powers even without regard for the advantage it brings has an unconditioned (moral) imperative rather than a conditioned (pragmatic) one as its basis, [391] can be shown in this way. The power to set an end—any end whatsoever—is the characteristic of humanity (as distinguished from animality). Hence there is also bound up with the end of humanity in our own person the rational will, and so the duty, to make ourselves worthy of humanity by culture in general, by procuring or promoting the *power* to realize all possible ends, so far as this power is to be found in man himself. In other words, man has a duty to cultivate the crude dispositions in human nature by which the animal first raises itself to man. To promote one's natural perfection is, accordingly, a duty in itself.

But this duty is merely ethical or of wide obligation. No principle of reason prescribes in a determinate way *how far* one should go in cultivating one's powers (in expanding or correcting one's power of understanding, i.e. in acquiring knowledge or skill). Then too, the different situations in which men may find themselves make a man's choice of the sort of occupation for which he should cultivate his talents quite arbitrary. With regard to natural perfection, accordingly, reason gives no law for actions but only a law for the maxims of actions, which runs as follows: "Cultivate your powers of mind and body so that they are fit to realize any end you can come upon," for it cannot be said which of these ends could, at some time, become yours.

b) The *cultivation of morality* in us. Man's greatest moral perfection is to do his duty and this *from a motive of duty* (to make the
law not merely the rule but also the motive of his actions).—Now
at first sight this looks like a *narrow* obligation, and the principle of
duty seems to prescribe with the precision and strictness of a law
not merely the *legality* but also the *morality* of every action, the
attitude of will. But in fact the law, here again, prescribes only the
maxim of the action: a maxim of seeking the ground of obligation
solely in the law and not in sensuous inclination (advantage or prej-
udice). It does not prescribe the *action itself.*—For man cannot so
scrutinize the depths of his own heart as to be quite certain, in even
a single action, of the purity of his moral purpose and the sincerity
of his attitude, even if he has no doubt about the legality of the
action. Very often he mistakes his own weakness, which counsels
him against the venture of a misdeed, for virtue (which is the notion
of strength); and how many people who have lived long and guiltless
lives [392] may not be merely *fortunate* in having escaped so many
temptations? It remains hidden from the agent himself how much
pure moral content there has been in the motive of each action.

Hence this duty too—the duty of valuing the worth of one's
actions not merely by their legality but also by their morality (our
attitude of will)—is of only wide obligation. The law does not prescribe this inner action in the human mind itself but only the
maxim of the action: the maxim of striving with all one's might to
make the thought of duty for its own sake the sufficient motive of
every dutiful action.

2. THE HAPPINESS OF OTHERS AS AN OBLIGATORY END

a) *Natural welfare.* Our *well-wishing* can be unlimited, since in it we
need do nothing. But *doing good* to others is harder, especially if we
should do it from duty, at the cost of sacrificing and mortifying
many of our desires, rather than from inclination (love) toward others.—The proof that beneficence is a duty follows from the fact that
our self-love cannot be divorced from our need of being loved by
others (i.e. of receiving help from them when we are in need), so
that we make ourselves an end for others. Now our maxim cannot
be obligatory [for others] unless it qualifies as a universal law and
so contains the will to make other men our ends too. The happiness
of others is, therefore, an end which is also a duty.

[The law says] only that I should sacrifice a part of my well-being to others without hope of requital, because this is a duty; it cannot assign determinate limits to the extent of this sacrifice. These limits will depend, in large part, on what a person's true needs consist of in view of his temperament, and it must be left to each to decide this for himself. For a maxim of promoting another's happiness at the sacrifice of my own happiness, my true needs, would contradict itself were it made a universal law.—Hence this duty is only a *wide* one: since no determinate limits can be assigned to what should be done, the duty has in it a play-room for doing more or less.—The law holds only for maxims, not for determinate actions. [393]

b) The happiness of another also includes his *moral well-being (salubritas moralis),* and we have a duty, but only a negative one, to promote this. Although the pain a man feels from the pangs of conscience has a moral source, it is still natural in its effect, like grief, fear, or any other morbid state. Now it is not *my* duty to prevent another person from deservedly experiencing this inner reproach: that is *his* affair. But it is my duty to refrain from anything that, considering the nature of men, could tempt him to do something for which his conscience would pain him; in other words, it is my duty not to give scandal.—But this concern for another's moral self-satisfaction does not admit of determinate limits being assigned to it, and so it is the ground of only a wide obligation.

IX. What is a Duty of Virtue?

Virtue is the strength of man's maxims in fulfilling his duty.—We can recognize strength of any kind only by the obstacles it can overcome, and in the case of virtue these obstacles are the natural inclinations, which can come into conflict with man's moral resolution. Now since it is man himself who puts these obstacles in the way of his maxims, virtue cannot be defined merely as self-constraint (for then he could use one natural inclination to restrain another). Virtue is, rather, self-constraint according to a principle of inner freedom, and so by the mere thought of one's duty in accordance with its formal law.

All duties contain a concept of *necessitation* by the law: *ethical* duties contain a necessitation such as can take place only in inner legislation—*juridical* duties, a necessitation such that outer legislation is also possible. Both, therefore, contain constraint, whether it be self-constraint or compulsion by another. And since the moral power of self-constraint can be called virtue, action springing from such an attitude (reverence for the law) can be called virtuous (ethical) action, even though the law asserts a juridical duty. For it is the *doctrine of virtue* that commands us to hold man's right holy.

But what it is virtuous to do is not necessarily a *duty of virtue* in the proper sense. The practice of virtue can have to do merely with the *formal* aspect of our maxims, while a duty of virtue is concerned with their matter—that is, with an [394] *end* which is also conceived as a duty.—Now because ethical obligation to ends, of which there can be many, contains merely a law for the maxims of actions and [goes beyond the formal condition of choice] to prescribe an end or matter (object) of choice, it is only *wide* obligation. So, just as there are different ends enjoined by the law, there are many different duties which we call *duties of virtue (officia honestatis)* because they admit of only free self-constraint, not compulsion by others, and because they determine ends which are also duties.

Like anything *formal,* virtue considered as the will's firm resolution to conform with every duty is always *one.* But in relation to the obligatory *end* of action or what one ought to make one's *end* (the material element in the maxim), there can be many virtues. And since obligation to the maxim of such an end is called a duty of virtue, there are many duties of virtue.

The first principle of the doctrine of virtue is: act according to a maxim of *ends* which it can be a universal law for everyone to have.—According to this principle man is an end, to himself as well as to others. And it is not enough that he has no title to use either himself or others merely as means (since according to this he can still be indifferent to them): it is in itself his duty to make man as such his end.

This first principle of the doctrine of virtue, as a categorical imperative, admits of no proof, but it does admit a deduction from pure practical reason.—What, in the relation of man to himself and others, *can be* an end, that *is* an end for pure practical reason. For pure practical reason is a power of ends as such, and for it to be indifferent to ends or to take no interest in them would be a contradiction, because then it would not determine the maxims for actions either (since every maxim contains an end) and so would not be practical reason. But pure reason can prescribe an end a priori only in so far as it declares it to be also a duty. And this duty is then called a duty of virtue. [395]

Notes

1. The deduction of the division of a system, i.e. the proof that the division is complete and final—which means that the transition from the concept divided to the entire series of the members of the division takes place without a leap *(divisio per saltum)*—is one of the most difficult conditions which the architect of a system has to fulfill. The question of what concept it is whose first division is that into right and wrong *(aut fas aut nefas)* should also give us pause. It is the act of free choice as such. But moral philosophers do not stop to reflect on this, even as the ontologist begins from something and nothing, as if they were the highest concepts, without seeing that they are already members of a division in which the concept divided is still wanting. This concept can be only that of an object as such [Kant's note].

2. Kant's sentence reads: "But because this act . . ." However, this sentence marks the transition from his discussion of action in general to his discussion of moral action in particular [translator's note].

3. *I.e.* "so far as actuality is concerned." The distinction is between actuality and potentiality. In other words, man's striving to raise himself from the crude state of his animality is not directed to his animality as a mere power for realizing ends, or potentiality, but rather to the actuality of these ends in human action [translator's note].

2

Utilitarianism and Moral Conflicts

John Stuart Mill

Logic: The Need for a Single Moral Principle

There is, then, a *philosophia prima* peculiar to Art, as there is one which belongs to Science. There are not only first principles of Knowledge, but first principles of Conduct. There must be some standard by which to determine the goodness or badness, absolute and comparative, of ends, or objects of desire. And whatever that standard is, there can be but one; for if there were several ultimate principles of conduct, the same conduct might be approved by one of those principles and condemned by another; and there would be needed some more general principle, as umpire between them.

Accordingly, writers on Moral Philosophy have mostly felt the necessity not only of referring all rules of conduct, and all judgments of praise and blame, to principles, but of referring them to some one

From *A System of Logic, Ratiocinative and Inductive* (New York: Harper and Brothers, 1884), book 6, chapter 12, section 7, and *Utilitarianism,* in *The Utilitarians* (Garden City, NY: Doubleday & Company, 1961), chapter 2, paragraph 25, and chapter 5, paragraphs 26–33 and 37–38. The subheads are provided by C.W.G.

principle; some rule, or standard, with which all other rules of conduct were required to be consistent, and from which by ultimate consequence they could all be deduced. Those who have dispensed with the assumption of such a universal standard, have only been enabled to do so by supposing that a moral sense, or instinct, inherent in our constitution, informs us, both what principles of conduct we are bound to observe, and also in what order these should be subordinated to one another.

The theory of the foundations of morality is a subject which it would be out of place, in a work like this, to discuss at large, and which could not to any useful purpose be treated incidentally. I shall content myself, therefore, with saying, that the doctrine of intuitive moral principles, even if true, would provide only for that portion of the field of conduct which is properly called moral. For the remainder of the practice of life some general principle, or standard, must still be sought; and if that principle be rightly chosen, it will be found, I apprehend, to serve quite as well for the ultimate principle of Morality, as for that of Prudence, Policy, or Taste.

Without attempting in this place to justify my opinion, or even to define the kind of justification which it admits of, I merely declare my conviction, that the general principle to which all rules of practice ought to conform, and the test by which they should be tried, is that of conduciveness to the happiness of mankind, or rather, of all sentient beings; in other words, that the promotion of happiness is the ultimate principle of Teleology.[1]

I do not mean to assert that the promotion of happiness should be itself the end of all actions, or even of all rules of action. It is the justification, and ought to be the controller, of all ends, but it is not itself the sole end. There are many virtuous actions, and even virtuous modes of action (though the cases are, I think, less frequent than is often supposed), by which happiness in the particular instance is sacrificed, more pain being produced than pleasure. But conduct of which this can be truly asserted, admits of justification only because it can be shown that, on the whole, more happiness will exist in the world, if feelings are cultivated which will make people, in certain cases, regardless of happiness. I fully admit that this is true; that the cultivation of an ideal nobleness of will and conduct should be to individual human beings an end, to which the specific pursuit either of their own happiness or of that of others (except so far as included in that idea) should, in any case of conflict, give way. But I hold that the very question, what constitutes *this elevation of character*, is itself to be decided by a reference to happiness as the standard. The character itself should be, to the individual, a paramount end, simply because the existence of this ideal nobleness of character, or of a near approach to it, in any abundance, would go farther than all things else toward making human life happy, both in the comparatively humble sense of pleasure and freedom from pain, and in the higher meaning, of rendering life, not what it now is almost universally, puerile and insignificant, but such as human beings with highly developed faculties can care to have.

.   .   .   .

Utilitarianism: The Resolution of Moral Conflicts

The remainder of the stock arguments against utilitarianism mostly consist in laying to its charge the common infirmities of human nature, and the general difficulties which embarrass conscientious persons in shaping their course through life. We are told that a utilitarian will be apt to make his own particular case an exception to moral rules, and when under temptation will see a utility in the breach of a rule greater than he will see in its observance. But is utility the only creed which is able to furnish us with excuses for evil-doing, and means of cheating our own conscience? They are afforded in abundance by all doctrines which recognize as a fact in morals the existence of conflicting considerations; which all doctrines do, that have been believed by sane persons. It is not the fault of any creed, but of the complicated nature of human affairs, that rules of conduct cannot be so framed as to require no exceptions, and that hardly any kind of action can safely be laid down as either always obligatory or always condemnable. There is no ethical creed which does not temper the rigidity of its laws by giving a certain latitude, under the moral responsibility of the agent, for accommodation to peculiarities of circumstances; and under every creed, at the opening thus made, self-deception and dishonest casuistry get in. There exists no moral system under which there do not arise unequivocal cases of conflicting obligation. These are the real difficulties, the knotty points both in the theory of ethics, and in the conscientious guidance of personal conduct. They are overcome practically, with greater or with less success, according to the intellect and virtue of the individual; but it can hardly be pretended that anyone will be the less qualified for dealing with them, from possessing an ultimate standard to which conflicting rights and duties can be referred. If utility is the ultimate source of moral obligations, utility may be invoked to decide between them when their demands are incompatible. Though the application of the standard may be difficult, it is better than none at all; while in other systems, the moral laws all claiming independent authority, there is no common umpire entitled to interfere between them: their claims to precedence one over another rest on little better than sophistry, and unless determined, as they generally are, by the unacknowledged influence of considerations of utility, afford a free scope for the action of personal desires and partialities. We must remember that only in these cases of conflict between secondary principles is it requisite that first principles should be appealed to. There is no case of moral obligation in which some secondary principle is not involved; and if only one, there can seldom be any real doubt which one it is, in the mind of any person by whom the principle itself is recognized.

. . . .

If the preceding analysis, or something resembling it, be not the correct account of the notion of justice; if justice be totally independent of utility, and be a standard per se, which the mind can recognize by simple introspection of itself; it is hard to understand why that internal oracle is so ambiguous, and why so many things appear either just or unjust, according to the light in which they are regarded.

We are continually informed that utility is an uncertain standard, which every different person interprets differently, and that there is no safety but in the immutable, ineffaceable, and unmistakable dictates of justice, which carry their evidence in themselves, and are independent of the fluctuations of opinion. One would suppose from this that on questions of justice there could be no controversy; that if we take that for our rule, its application to any given case could leave us in as little doubt as a mathematical demonstration. So far is this from being the fact, that there is as much difference of opinion and as much discussion about what is just, as about what is useful to society. Not only have different nations and individuals different notions of justice, but in the mind of one and the same individual, justice is not some one rule, principle, or maxim, but many, which do not always coincide in their dictates, and in choosing between which he is guided either by some extraneous standard or by his own personal predilections.

For instance, there are some who say that it is unjust to punish anyone for the sake of example to others; that punishment is just only when intended for the good of the sufferer himself. Others maintain the extreme reverse, contending that to punish persons who have attained years of discretion, for their own benefit, is despotism and injustice, since if the matter at issue is solely their own good, no one has a right to control their own judgment of it; but that they may justly be punished to prevent evil to others, this being the exercise of the legitimate right of self-defense. Mr. Owen,[2] again, affirms that it is unjust to punish at all; for the criminal did not make his own character; his education, and the circumstances which surrounded him, have made him a criminal, and for these he is not responsible. All these opinions are extremely plausible; and so long as the question is argued as one of justice simply, without going down to the principles which lie under justice and are the source of its authority, I am unable to see how any of these reasoners can be refuted. For in truth every one of the three builds upon rules of justice confessedly true. The first appeals to the acknowledged injustice of singling out an individual, and making him a sacrifice, without his consent, for other people's benefit. The second relies on the acknowledged justice of self-defense, and the admitted injustice of forcing one person to conform to another's notions of what constitutes his good. The Owenite invokes the admitted principle that it is unjust to punish anyone for what he cannot help. Each is triumphant so long as he is not compelled to take into consideration any other maxims of justice than the one he has selected; but as soon as their several maxims are brought face to face, each disputant seems to have exactly as much to say for himself as the others. No one of them can carry out his own notion of justice without trampling upon another equally binding. These are difficulties; they have always been felt to be such; and many devices have been invented to turn rather than to overcome them. As a refuge from the last of the three, men imagined what they called the freedom of the will; fancying that they could not justify punishing a man whose will is in a thoroughly hateful state, unless it be supposed to have come into that state through no influence of anterior circumstances. To escape from the other difficulties, a favorite contrivance has been the fiction of a contract, whereby at some unknown period all the members of society engaged to obey the laws, and consented to be punished for any disobedience to them; thereby giving to their legislators the right, which it is assumed they would not otherwise have had, of punishing them, either for their own good or for that of society. This happy thought was considered to get rid of the whole difficulty, and to legitimate the infliction of punishment, in virtue of another received maxim of justice, *Volenti non fit injuria*—that is not unjust which is done with the consent of the person who is supposed to be hurt by it. I need hardly remark that even if the consent were not a mere fiction, this maxim is not superior in authority to the others which it is brought in to supersede. It is, on the contrary, an instructive specimen of the loose and irregular manner in which supposed principles of justice grow up. This particular one evidently came into use as a help to the coarse exigencies of courts of law, which are sometimes obliged to be content with very uncertain presumptions, on account of the greater evils which would often arise from any attempt on their part to cut finer. But even courts of law are not able to adhere consistently to the maxim, for they allow voluntary engagements to be set aside on the ground of fraud, and sometimes on that of mere mistake or misinformation.

Again, when the legitimacy of inflicting punishment is admitted, how many conflicting conceptions of justice come to light in discussing the proper apportionment of punishments to offenses. No rule on the subject recommends itself so strongly to the primitive and spontaneous sentiment of justice as the *lex talionis,* an eye for an eye and a tooth for a tooth. Though this principle of the Jewish and of the Mohammedan law has been generally abandoned in Europe as a practical maxim, there is, I suspect, in most minds, a secret hankering after it; and when retribution accidentally falls on an offender in that precise shape, the general feeling of satisfaction evinced bears witness how natural is the sentiment to which this repayment in kind is acceptable. With many, the test of justice in penal infliction is that the punishment should be proportioned to the offense; meaning that it should be exactly measured by the moral guilt of the culprit (whatever be their standard for measuring moral guilt)—the consideration, what amount of punishment is necessary to deter from the offense, having nothing to do with the question of justice, in their estimation; while there are others to whom that consideration is all in all—who maintain that it is not just, at least for man, to inflict on a fellow-creature, whatever may be his offenses, any amount of suffering beyond the least that will suffice to prevent him from repeating, and others from imitating, his misconduct.

To take another example from a subject already once referred to. In a co-operative industrial association, is it just or not that talent or skill should give a title to superior remuneration? On the negative side of the question it is argued that whoever does the best he can, deserves equally well, and ought not in justice to be put in a position of inferiority for no fault of his own; that superior abilities have already advantages more than enough, in the admiration they excite, the personal influence they command, and the internal sources of satisfaction attending them, without adding to these a superior share of the world's goods; and that society is bound in justice rather to make compensation to the less favored for this unmerited inequality of advantages, than to aggravate it. On the contrary side it is contended that society receives more from the more efficient laborer; that his services being more useful, society owes him a larger return for them; that a greater share of the joint result is actually his work, and not to allow his claim to it is a kind of robbery; that if he is only to receive as much as others, he can only be justly required to produce as much, and to give a smaller amount of time and exertion, proportioned to his superior efficiency. Who shall decide between these appeals to conflicting principles of justice? Justice has in this case two sides to it, which it is impossible to bring into harmony, and the two disputants have chosen opposite sides; the one looks to what it is just that the individual should receive, the other to what it is just that the community should give. Each, from his own point of view, is unanswerable; and any choice between them, on grounds of justice, must be perfectly arbitrary. Social utility alone can decide the preference.

How many, again, and how irreconcilable, are the standards of justice to which reference is made in discussing the repartition of taxation. One opinion is, that payment to the State should be in numerical proportion to pecuniary means. Others think that justice dictates what they term graduated taxation; taking a higher percentage from those who have more to spare. In point of natural justice a strong case might be made for disregarding means altogether, and taking the same absolute sum (whenever it could be got) from everyone—as the subscribers to a mess, or to a club, all pay the same sum for the same privileges, whether they can all equally afford it or not. Since the protection (it might be said) of law and government is afforded to, and is equally required by all, there is no injustice in making all buy it at the same price. It is reckoned justice, not injustice, that a dealer should charge to all customers the same price for the same article, not a price varying according to their means of payment. This doctrine, as applied to taxation, finds no advocates, because it conflicts so strongly with man's feelings of humanity and of social expediency; but the principle of justice which it invokes is as true and as binding as those which can be appealed to against it. Accordingly it exerts a tacit influence on the line of defense employed for other modes of assessing taxation. People feel obliged to argue that the State does more for the rich than for the poor, as a justification for its taking more from them: though this is in reality not true, for the rich would be far better able to protect themselves, in the absence of law or government, than the poor, and indeed would probably be successful in converting the poor into their slaves. Others, again, so far defer to the same conception of justice, as to maintain that all should pay an equal capitation tax for the protection of their persons (these being of equal value to all), and an unequal tax for the protection of their property, which is unequal. To this others reply that the all of one man is as valuable to him as the all of another. From these confusions there is no other mode of extrication than the utilitarian.

Is, then, the difference between the just and the expedient a merely imaginary distinction? Have mankind been under a delusion in thinking that justice is a more sacred thing than policy, and that the latter ought only to be listened to after the former has been satisfied? By no means. The exposition we have given of the nature and origin of the sentiment, recognizes a real distinction; and no one of those who profess the most sublime contempt for the consequences of actions as an element in their morality, attaches more importance to the distinction than I do. While I dispute the pretensions of any theory which sets up an imaginary standard of justice not grounded on utility, I account the justice which is grounded on utility to be the chief part, and incomparably the most sacred and binding part, of all morality. Justice is a name for certain classes of moral rules which concern the essentials of human well-being more nearly, and are therefore of more absolute obligation, than any other rules for the guidance of life; and the notion which we have found to be of the essence of the idea of justice, that of a right residing in an individual, implies and testifies to this more binding obligation.

The moral rules which forbid mankind to hurt one another (in which we must never forget to include wrongful interference with each other's freedom) are more vital to human well-being than any maxims, however important, which only point out the best mode of managing some department of human affairs. They have also the peculiarity, that they are the main element in determining the whole of the social feelings of mankind. It is their observance which alone preserves peace among human beings: if obedience to them were not the rule, and disobedience the exception, everyone would see in everyone else an enemy, against whom he must be perpetually guarding himself. What is hardly less important, these are the precepts which mankind have the strongest and the most direct inducements for impressing upon one another. By merely giving to each other prudential instruction or exhortation, they may gain, or think they gain, nothing; in inculcating on each other the duty of positive beneficence they have an unmistakable interest, but far less in degree: a person may possibly not need the benefits of others, but he always needs that they should not do him hurt. Thus the moralities which protect every individual from being harmed by others, either directly or by being hindered in his freedom of pursuing his own good, are at once those which he himself has most at heart, and those which he has the strongest interest in publishing and enforcing by word and deed. It is by a person's observance of these that his fitness to exist as one of the fellowship of human beings is tested and decided; for on that depends his being a nuisance or not to those with whom he is in contact. Now it is these moralities primarily which compose the obligations of justice. The most marked cases of injustice, and those which give the tone to the feeling of repugnance which characterizes the sentiment, are acts of wrongful aggression, or wrongful exercise of power over someone; the next are those which consist in wrongfully withholding from him something which is his due: in both cases, inflicting on him a positive hurt, either in the form of direct suffering, or of the privation of some good which he had reasonable ground, either of a physical or of a social kind, for counting upon.

. . . .

It appears from what has been said that justice is a name for certain moral requirements which, regarded collectively, stand higher in the scale of social utility, and are therefore of more paramount obligation, than any others; though particular cases may occur in which some other social duty is so important, as to overrule any one of the general maxims of justice. Thus, to save a life, it may not only be allowable but a duty to steal or take by force the necessary food or medicine, or to kidnap and compel to officiate the only qualified medical practitioner. In such cases, as we do not call anything justice which is not a virtue; we usually say, not that justice must give way to some other moral principle, but that what is just in ordinary cases is, by reason of that other principle, not just in the particular case. By this useful accommodation of language, the character of indefeasibility attributed to justice is kept up, and we are saved from the necessity of maintaining that there can be laudable injustice.

The considerations which have now been adduced resolve, I conceive, the only real difficulty in the utilitarian theory of morals. It has always been evident that all cases of justice are also cases of expediency: the difference is in the peculiar sentiment which attaches to the former, as contradistinguished from the latter. If this characteristic sentiment has been sufficiently accounted for; if there is no necessity to assume for it any peculiarity of origin; if it is simply the natural feeling of resentment, moralized by being made coextensive with the demands of social good; and if this feeling not only does but ought to exist in all the classes of cases to which the idea of justice corresponds: that idea no longer presents itself as a stumbling-block to the utilitarian ethics. Justice remains the appropriate name for certain social utilities which are vastly more important, and therefore more absolute and imperative, than any others are as a class (though not more so than others may be in particular cases); and which, therefore, ought to be, as well as naturally are, guarded by a sentiment not only different in degree, but also in kind; distinguished from the milder feeling which attaches to the mere idea of promoting human pleasure or convenience, at once by the more definite nature of its commands, and by the sterner character of its sanctions.

Notes

1. For an express discussion and vindication of this principle, see the little volume entitled *Utilitarianism*.
2. Robert Owen, a nineteenth century British author and reformer, was a forerunner of Socialism and founded a cooperative community in Indiana named New Harmony. Among his philosophical views was the claim that a person's character is formed by his circumstances.—C.W.G.

3

Collision of Duties

F. H. Bradley

Collision of Duties Unavoidable

Thus[1] to get from the form of duty to particular duties is impossible.
The particular duties must be taken for granted, as in ordinary
morality they are taken for granted. But supposing this done, is duty
for duty's sake a valid formula, in the sense that we are to act always
on a law and nothing but a law, and that a law can have no exceptions, in the sense of particular cases where it is overruled? No, this
takes for granted that life is so simple that we never have to consider
more than one duty at a time; whereas we really have to do with
conflicting duties, which as a rule escape conflict simply because it
is understood which have to give way. It is a mistake to suppose

From the second edition of *Ethical Studies* (Oxford: Oxford University Press, 1927),
pp. 156–59, 193–99, and 214–28. The title is provided by C.W.G. Headings are taken
from the table of contents; notes in square brackets appeared in the second edition,
based on unfinished notes Bradley had prepared before his death. References in the
text and the notes are to *Ethical Studies*.

that collision of duties is uncommon; it has been remarked truly that *every* act can be taken to involve such collision.[2]

To put the question plainly—It is clear that in a given case I may have several duties, and that I may be able to do only one. I must then break some 'categorical' law, and the question the ordinary man puts to himself is, Which duty am I to do? He would say, 'All duties have their limits and are subordinated one to another. You can not put them all in the form of your "categorical imperative" (in the shape of a law absolute and dependent on nothing besides itself) without such exceptions and modifications that, in many cases, you might as well have left it alone altogether. We certainly have laws, but we may not be able to follow them all at once; and to know which we are to follow is a matter of good sense which can not be decided in any other way. One should give to the poor—in what cases and how much? Should sacrifice oneself—in what way and within what limits? Should not indulge one's appetites—except when it is right. Should not idle away one's time—except when one takes one's pleasure. Nor neglect one's work—but for some good reason. All these points we admit are in one way matter of law; but if you think to decide in particular cases by applying some "categorical imperative," you must be a pedant, if not a fool.'

Ordinary morality does not hold to each of its laws as inviolable, each as an absolute end in itself. It is not even aware of a collision in most cases where duties clash; and, where it perceives it, and is confronted with collisions of moral laws, each of which it has been accustomed to look on as an absolute monarch, so to speak, or a commander-in-chief, rather than as a possible subordinate officer, there it does subordinate one to the other, and feels uneasiness only in proportion to the rarity of the necessity, and the consequent jar to the feelings. There are few laws a breach of which (in obedience to a higher law) morality does not allow, and I believe there is none which is not to be broken in conceivable (imaginable) circumstances,[3] though the necessity of deciding the question does not practically occur. According to ordinary morality (the fact is too palpable to be gainsaid), it is quite right to speak falsely with intent to deceive under certain circumstances, though ordinary morality might add, 'I don't call that a lie'. It *is* a lie; and when Kant and others maintained that it must always be wrong to lie, they forgot the rather important fact that in some cases to abstain from acting *is* acting, is wilful neglect of a duty, and that there are duties above truth-speaking, and many offences against morality which are worse, though they may be less painful, than a lie. So to kill oneself, in a manner which must be called suicide, *may* not only be right but heroic[4]; homicide may be excusable, rebellion in the subject and disobedience in the soldier all morally justifiable, and every one of them clear breaches of categorical imperatives, in obedience to a higher law.

All that it comes to is this (and it is, we must remember, a very important truth), that you must never break a law of duty to please yourself, never for the sake of an end not duty, but only for the sake of a superior and overruling duty. Any breach of duty, as duty, and not as *lower* duty, is always and absolutely wrong; but it would be rash to say that any one act must be in all cases absolutely and unconditionally immoral. Circumstances decide, because circumstances determine the manner in which the overruling duty must be realized. This is a simple fact which by the candid observer can not be denied, and which is merely the exposition of the moral consciousness, though I am fully aware that it is an exposition which that consciousness would not accept, simply because it must necessarily misunderstand it in its abstract form. And if moral theory were meant to influence moral practice and to be dabbled in by 'the vulgar' (and there are not so many persons who in this respect are *not* the vulgar), then I grant this is a fact it would be well to keep in the background. None the less it is a fact.[5]

So we see 'duty for duty's sake' says only, 'do the right for the sake of the right'; it does not tell us what right is; or 'realize a good will, do what a good will would do, for the sake of being yourself a good will'. And that is something; but beyond that it is silent or beside the mark. It tells us to act for the sake of a form, which we saw was a self-contradictory command; and we even saw that in sober sadness the form did exist for form's sake, and in literal truth remained only a form. We saw that duty's universal laws are not universal, if that means they can never be overruled, and that its form and its absolute imperative are impracticable. What after all remains is the acting for the sake of a good will, to realize oneself by realizing the will which is above us and higher than ours, and the assurance that this, and not the self to be pleased, is the end for which we have to live. But as to that which the good will is it tells us nothing, and leaves us with an idle abstraction.

.   .   .   .

Intuitive Character of Moral Judgements

The next point we come to is the question, How do I get to know in particular what is right and wrong? And here again we find a strangely erroneous preconception. It is thought that moral philosophy has to accomplish this task for us; and the conclusion lies near at hand, that any system which will not do this is worthless. Well, we first remark, and with some confidence, that there cannot be a moral philosophy which will tell us what in particular we are to do, and also that it is not the business of philosophy to do so. All philosophy has to do is 'to understand what is', and moral philosophy has to understand morals which exist, not to make them or give directions for making them. Such a notion is simply ludicrous. Philosophy in general has not to anticipate the discoveries of the particular sciences nor the evolution of history; the philosophy of religion has not to make a new religion or teach an old one, but simply to understand the religious consciousness; and aesthetic has not to produce works of fine art, but to theorize the beautiful which it finds; political philosophy has not to play tricks with the state, but to understand it; and ethics has not to make the world moral, but to reduce to theory the morality current in the world. If we want it to do anything more, so much the worse for us; for it can not possibly construct new morality, and, even if it could to any extent codify what exists (a point on which I do not enter), yet it surely is clear that in cases of collision of duties it would not help you to know what to do. Who would go to a learned theologian, as such, in a practical religious difficulty; to a system of aesthetic for suggestions on the handling of an artistic theme; to a physiologist, as such, for a diagnosis and prescription; to a political philosopher in practical politics; or to a psychologist in an intrigue of any kind? All these persons no doubt *might* be the best to go to, but that would not be because they were the best theorists, but because they were more. In short, the view which thinks moral philosophy is to supply us with particular moral prescriptions confuses science with art, and confuses, besides, reflective with intuitive judgement. That which tells us what in particular is right and wrong is not reflection but intuition.[6]

We know what is right in a particular case by what we may call an immediate judgement, or an intuitive subsumption. These phrases are perhaps not very luminous, and the matter of the 'intuitive understanding' in general is doubtless difficult, and the special character of moral judgements not easy to define; and I do not say that I am in a position to explain these subjects at all, nor, I think, could any one do so, except at considerable length. But the point that I do wish to establish here is, I think, not at all obscure. The reader has first to recognize that moral judgements are not discursive; next, that nevertheless they do start from and rest on a certain basis; and then if he puts the two together, he will see that they involve what he may call the 'intuitive understanding', or by any other name, so long as he keeps in sight the two elements and holds them together.

On the head that moral judgments are not discursive, no one, I think, will wish me to stay long. If the reader attends to the facts he will not want anything else; and if he does not, I confess I can not prove my point. In practical morality no doubt we *may* reflect on our principles, but I think it is not too much to say that we *never* do so, except where we have come upon a difficulty of particular application. If any one thinks that a man's *ordinary* judgement, 'this is right or wrong,' comes from the having a rule *before* the mind and bringing the particular case under it, he may be right; and I can not try to show that he is wrong. I can only leave it to the reader to judge for himself. We say we 'see' and we 'feel' in these cases, not we 'conclude'. We prize the advice of persons who can give us no reasons for what they say. There is a general belief that the having a reason for all your actions is pedantic and absurd. There is a general belief that to try to have reasons for all that you do is sometimes very dangerous. Not only the woman but the man who 'deliberates' may be 'lost'. First thoughts are often the best,[7] and if once you begin to argue with the devil you are in a perilous state. And I think I may add (though I do it in fear) that women in general are remarkable for the fineness of their moral perceptions[8] and the quickness of their judgements, and yet are or (let me save myself by saying) 'may be' not remarkable for corresponding discursive ability.

Taking for granted then that our ordinary way of judging in morals is not by reflection and explicit reasoning, we have now to point to the other side of the fact, viz. that these judgements are not mere isolated impressions, but stand in an intimate and vital relation to a certain system, which is their basis. Here again we must ask the reader to pause, if in doubt, and consider the facts for himself. Different men, who have lived in different times and countries, judge or would judge a fresh case in morals differently. Why is this? There is probably no 'why' before the mind of either when he judges; but *we* perhaps can say, 'I know why A said so and B so', because we find some general rule or principle different in each, and in each the basis of the judgement. Different people in the same society may judge points differently, and we sometimes know why. It is because A is struck by one aspect of the case, B by another; and one principle is (not *before,* but) *in* A's mind when he judges, and another in B's. Each has subsumed, but under a different head; the one perhaps justice, the other gratitude. Every man has the morality he has made his own in his mind, and he 'sees' or 'feels' or 'judges' accordingly, though he does not reason explicitly from data to a conclusion.

I think this will be clear to the reader; and so we must say that on their perceptive or intellectual side (and that, the reader must not forget, is the one side that we are considering) our moral judgements are intuitive subsumptions.

To the question, How am I to know what is right? the answer must be, By the $\alpha'\iota\sigma\theta\eta\sigma\iota\varsigma$ of the $\phi\rho\acute{o}\nu\iota\mu o\varsigma$; and the $\phi\rho\acute{o}\nu\iota\mu o\varsigma$ is the man who has identified his will with the moral spirit of the community, and judges accordingly. If an immoral course be suggested to him, he 'feels' or 'sees' at once that the act is not in harmony with a good will, and he does not do this by saying, 'this is a breach of rule A, *therefore,* etc.'; but the first thing he is aware of is that he 'does not like it'; and what he has done, without being aware of it, is (at least in most cases) to seize the quality of the act, that quality being a general quality. Actions of a particular kind he does not like, and he has instinctively referred the particular act to that kind. What is right is perceived in the same way; courses suggest themselves, and one is approved of, because intuitively judged to be of a certain kind, which kind represents a principle of the good will.

If a man is to know what is right, he should have imbibed by precept, and still more by example, the spirit of his community, its general and special beliefs as to right and wrong, and, with this whole embodied in his mind, should particularize it in any new case, not by a reflective deduction, but by an intuitive subsumption, which does not know that it is a subsumption[9]; by a carrying out of the self into a new case, wherein what is before the mind is the case and not the self to be carried out, and where it is indeed the whole that feels and sees, but all that is seen is seen in the form of *this* case, *this* point, *this* instance. Precept is good, but example is better; for by a series of particulars (as such forgotten) we get the general spirit, we identify ourselves on the sides both of will and judgement with the basis, which basis (be it remembered) has not got to be explicit.[10]

There are a number of questions which invite consideration[11] here, but we can not stop. We wished to point out briefly the character of our common moral judgements. This (on the intellectual side) is the way in which they are ordinarily made; and, in the main, there is not much practical difficulty. What is moral *in any particular given case* is seldom doubtful.[12] Society pronounces beforehand[13]; or, after some one course has been taken, it can say whether it was right or not; though society can not generalize much, and, if asked to reflect, is helpless and becomes incoherent. But I do not say there are no cases where the morally-minded man has to doubt; most certainly such do arise, though not so many as some people think, far fewer than some would be glad to think. A very large number arise from reflection, which wants to act from an explicit principle, and so begins to abstract and divide, and, thus becoming one-sided, makes the relative absolute. Apart from this, however, collisions must take place; and here there is no guide whatever but the intuitive judgement of oneself or others.[14]

This intuition must not be confounded with what is sometimes miscalled 'conscience'. It is not mere individual opinion or caprice. It presupposes the morality of the community as its basis, and is subject to the approval thereof. Here, if anywhere, the idea of universal and impersonal morality is realized. For the final arbiters are the φρόνιμοι, persons with a will to do right, and not full of reflections and theories. If they fail you, you must judge for yourself, but practically they seldom do fail you. Their private peculiarities neutralize each other, and the result is an intuition which does not belong merely to this or that man or collection of men. 'Conscience' is the antipodes of this. It wants you to have no law but yourself, and to be better than the world. But this intuition tells you that, if you could be as good as your world, you would be better than most likely you are, and that to wish to be better than the world is to be already on the threshold of immorality.

. . . .

No Limit to the Moral Sphere

In our criticism of [the doctrine of "my station and its duties"] developed in Essay V [of *Ethical Studies*] we saw that, however true the main doctrine of that Essay may be, it is no sufficient answer to the question, What is morality? and, guided by its partial failure, we must try to find a less one-sided solution.

We saw (in Essay II [of *Ethical Studies*]) that the end was the realizing of the self; and the problem which in passing suggested itself was, Are morality and self-realization the same thing,[15] or, if not altogether the same, in what respect are they different?

That in some way they do differ is clear from the popular views on the subject. Every one would agree that by his artistic or scientific production an artist or a man of science does realize himself, but no one, not blinded by a theory, would say that he was moral just so far as, and because, what he produced was good of its sort and desirable in itself. A man may be good at this or that thing, and may have done good work in the world; and yet when asked, 'But was he a good man?' we may find ourselves, although we wish to say Yes, unable to do more than hesitate. A man need not be a good man just so far as he is a good artist; and the doctrine which unreservedly identifies moral goodness with any desirable realization of the self can not be maintained.

Can we then accept the other view, which, as it were, separates morality into a sphere of its own; which calls a man moral according as he abstains from direct breaches of social rules, and immoral if he commits them; while it forgets that the one man may be lazy, selfish, and without a wish to improve himself, while the other, with all his faults, at least loves what is beautiful and good, and has striven towards it? We can not do that unless, while we recognize the truth of the doctrine, we shut our eyes to its accompanying falsity.

And, finding in neither the expression of our moral consciousness, we thankfully accept the correction which sees in 'conduct' nine-tenths of life, though we can not expect the main question to be answered by a coarse and popular method, which divides into parts instead of distinguishing aspects; and though, in the saving one-tenth and the sweeping nine-tenths alike, we can see little more than the faltering assertion of one mistake, or the confident aggravation of another.

A man's life, we take it, can not thus be cut in pieces. You can not say, 'In this part the man is a moral being, and in that part he is not'. We have not yet found that fraction of his existence in which the moral goodness of the good man is no more realized, and where 'the lusts of the flesh' cease to wage their warfare. We have heard in the sphere of religion, 'Do *all* to the glory of God', and here too we recognize no smaller claim. To be a good man in all things and everywhere, to try to do always the best, and to do one's best in it, whether in lonely work or in social relaxation to suppress the worse self and realize the good self, this and nothing short of this is the dictate of morality. This, it seems to us, is a deliverance of the moral consciousness too clear for misunderstanding, were it not for two fixed habits of thought. One of these lies in the confining of a man's morality to the sphere of his social relations; the other is the notion that morality is a life harassed and persecuted everywhere by 'imperatives' and disagreeable duties, and that without these you have not got morality. We have seen, and have yet to see, that the first has grasped only part of the truth; and on the second it is sufficient to remark that it stands and falls with the identification of morality with unwilling obedience to law, and that, according to the common view, a man does not cease to be good so far as goodness becomes natural and pleasant to him.

But we shall be met at this point with an absurdity supposed to follow. Work of any sort, it will be said, is, we grant you, a field for morality, and so is most of life in relation to others; but there must be a sphere where morality ceases, or else it will follow that a man is moral in all the trifling details of his own life which concern him alone, and no less again in his amusements. If morality does not stop somewhere, you must take it to be a moral question not only whether a man amuses himself, but also how he amuses himself. There will be no region of things indifferent, and this leads to consequences equally absurd and immoral.[16] We answer without hesitation that in human life there is, in one sense, no sphere of things indifferent, and yet that no absurd consequences follow. If it is my moral duty to go from one town to another, and there are two roads which are equally good, it is indifferent to the proposed moral duty *which* road I take; it is not indifferent *that* I do take one or the other; and whichever road I do take, I am doing my duty on it, and hence it is far from indifferent: my walking on road A is a matter of duty in reference to the end, though not a matter of duty if you consider it against walking on road B; and so with B—but I can escape the sphere of duty neither on A nor on B. In order to realize the good will in a finite corporeal being it is necessary that certain spheres should exist, and should have a general character; this is a moral question, and not indifferent. The detail of those spheres within certain limits does not matter; not that it is immaterial that there *is* a detail of trifles, and hence not that this and that trifle has no moral importance, but that this trifle has no importance *against that trifle.* Qualify a trifle by subordinating it to a good will, and it has moral significance; qualify it by contrast with another trifle, and morally it signifies nothing. This is plain enough, and, so far as it goes, will I hope be sufficient. The reader no doubt will see that, if a class of acts is morally desirable, then whatever falls within that class is also morally desirable, so far as falling therein; though in its other relations it may be indifferent.

But the difficulty which remains will be something of this sort. The reader will feel that, to a certain extent, the regulation of the times and fields of amusements, etc., and, to a still larger extent, the choice of trifling details therein, involves no reflection, no deliberate choice, is not made a matter of conscience, is in a word done naturally; and he may find a difficulty in seeing how, if this is so, it can be said to fall within the moral sphere. Morality, he may feel, does tell me it is good to amuse myself, and more decidedly that I may *not* amuse myself beyond certain limits; but within those limits it leaves me to my natural self. In this, it seems to us, there is a twofold misapprehension, a mistake as to the limits, and a mistake as to the character of the moralized self. It is, first, an error to suppose that in what is called human life there remains any region which has not been moralized. Whatever has been brought under the control of the will, it is not too much to say, has been brought into the sphere of morality; in our eating, our drinking, our sleeping, we from childhood have not been left to ourselves; and the habits, formed in us by the morality outside of us, now hold of the moral will which in a manner has been their issue. And so in our lightest moments the element of control and regulation is not wanting; it is part of the business of education to see that it is there, and its absence, wherever it is seen to be absent, pains us. The character shows itself in every trifling detail of life; we can not go in to amuse ourselves while we leave it outside the door with our dog; it is ourself, and our moral self, being not mere temper or inborn disposition, but the outcome of a series of acts of will. Natural it is indeed well to be; but that is because by this time morality should be our nature, and good behaviour its unreflecting issue; and to be natural in any sense which excludes moral habituation is never, so far as I know the world, thought desirable. In a good and amiable man the good and amiable self is present throughout, and that self is for us a moral self. This brings us to the second mistake, which also rests on the same misapprehension of the cardinal truth that what is natural can not be moral, nor what is moral natural. 'What is natural does not reflect, and without reflection there is no morality. Hence, where we are natural *because* we do not reflect, there we can not be moral'. So runs the perversion. But here it is forgotten that we *have* reflected; that acts which issue from moral reflection have qualified our will; that our character thus, not only in its content, but also in the form of its acquisition, is within the moral sphere; and that a character, whether good or bad, is a second nature. The man to whom it 'comes natural' to be good is commonly thought a good man, and the good self of the good man is present in and determines the detail of his life not less effectually because unconsciously. So far facts speak loudly, and the only path which remains open to the objector is to deny that the good self is necessarily a moral self, on the ground not that its content is nonmoral, but that its genesis is so; in other words, because, though moral in itself, it is not so for the agent. We may be told, the genesis of the good self generally is not a moral genesis, or in this and that sphere or relation it is not so, and hence, though good, it need not, so far as good, be moral. To the consideration of this question we shall have to come later, and at present can only observe that we refuse to separate goodness conscious or unconscious from the will to be good, or the will to be good from morality; and we assert that, because the good self shows itself everywhere, therefore there is no part of life at which morality stops and goes no further. Thus much against the notion that in our amusements, etc., we cease to be moral beings, that there is a tenth part of life where conduct is not required. But as to the remaining nine-tenths we need surely say no more: wherever there is anything to be done not in play but in earnest, there the moral consciousness tells us it is right to do our best, and, if this is so, there can be no question but that here is a field for morality.[17]

It is a moral duty to realize everywhere the best self, which for us in this sphere is an ideal self; and, asking what morality is, we so far must answer, it is coextensive with self-realization in the sense of the realization of the ideal self in and by us. And thus we are led to the inquiry, what is the *content* of this ideal self.[18]

Content of the Ideal Self

From our criticism on the foregoing Essay we can at once gather that
the good self is the self which realizes (1) a social, (2) a nonsocial
ideal; the self, first, which does, and, second, which does not directly
and immediately involve relation to others. Or from another point
of view, what is aimed at is the realization in me (1) of the ideal
which is realized in society, of my station and its duties, or (2) of
the ideal which is not there fully realized; and this is (*a*) the perfection of a social and (*b*) of a nonsocial self. Or again (it is all the same
thing) we may divide into (1) duties to oneself which are not
regarded as social duties, (2) duties to oneself[19] which are so
regarded, these latter being (*a*) the duties of the station which I happen to be in, (*b*) duties beyond that station. Let us further explain.

The content of the good self, we see, has a threefold origin; and
(1) the first and most important contribution comes from what we
have called my station and its duties, and of this we have spoken
already at some length. We saw [in Essay V of *Ethical Studies*] that
the notion of an individual man existing in his own right independent of society was an idle fancy, that a human being is human
because he has drawn his being from human society, because he is
the individual embodiment of a larger life; and we saw that this
larger life, of the family, society, or the nation, was a moral will, a
universal the realization of which in his personal will made a man's
morality. We have nothing to add here except in passing to call
attention to what we lately advanced, viz. that the good man is good
throughout all his life and not merely in parts; and further to request
the reader to turn to himself and ask himself in what his better self
consists. He will find, if we do not mistake, that the greater part of
it consists in his loyally, and according to the spirit, performing his
duties and filling his place as the member of a family, society, and
the state. He will find that, when he has satisfied the demands of
these spheres upon him, he will in the main have covered the claims
of what he calls his good self. The basis and foundation of the ideal
self is the self which is true to my station and its duties.

But (2) we saw also that, if we investigate our good self, we find
something besides, claims beyond what the world expects of us, a
will for good beyond what we see to be realized anywhere. The good
in my station and its duties was visibly realized in the world, and it
was mostly possible to act up to that real ideal; but this good beyond

is only an ideal; for it is not wholly realized in the world we see, and, do what we may, we can not find it realized in ourselves. It is what we strive for and in a manner do gain, but never attain to and never possess. And this ideal self (so far as we are concerned with it here) is a social self. The perfect types of zeal and purity, honour and love, which, figured and presented in our own situation and circumstances, and thereby unconsciously specialized, become the guides of our conduct and law of our being, are social ideals. They directly involve relation to other men, and, if you remove others, you immediately make the practice of these virtues impossible.[20]

This then is the ideal self which in its essence is social; and concerning this many difficulties arise which we can not discuss. Among these would be the two inquiries, What is the origin, and what the content of this ideal self? In passing we may remark that the first contains two questions which are often confused, viz. (a) How is it possible for the mind to frame an ideal; or, given as a fact a mind which idealizes, what must be concluded as to its nature? Can anything idealize unless itself in some way be an ideal? This, we need not say, suggests serious problems which we can not even touch upon here. Then (b) it contains also the questions, What was the historical genesis of the ideal; by what steps did it come into the world? And again, What is its genesis in us? And these can scarcely be separated from one another, or from the further inquiry, What is its content?

The historical genesis we shall not enter on; and as to the genesis in the individual, we will merely remark that we seem first to see in some person or persons the type of what is excellent; then by the teaching, tradition, and imagination of our own and other countries and times, we receive a content which we find existing realized in present or past individuals, and finally detach from all as that which is realized wholly in none, but is an ideal type of human perfection. At this point we encounter a question of fact, namely, how far the ideal which serves as a guide to conduct is presented in an individual form. No doubt two extremes exist. A large number of men have, I think, no moral ideal beyond the station they live in, and of these some are even satisfied with the presentation of this or that known person as a type; while again in the highest form of morality the ideal is not figured in the shape of an individual.[21] But between the extremes must be endless gradations.

We have previously said something as to the way in which the ideal is made use of in moral judgements, and what remains is to call attention to the content of this social ideal. It is obvious at once that it is a will which practises no other kind of virtues than those which we find in the world; and we can see no reason for supposing this presented ideal self to be anything beyond the idealization of what exists in human nature, the material idealized being more or less cosmopolitan, and the abstraction employed being more or less one-sided.

And with these cursory and insufficient remarks we must dismiss the ideal of a perfect social being.

But (3) there remains in the good self a further region we have not yet entered on; an ideal, the realization of which is recognized as a moral duty, but which yet in its essence does not involve direct relation to other men.[22] The realization for myself of truth and beauty, the living for the self which in the apprehension, the knowledge, the sight, and the love of them finds its true being, is (all those who know the meaning of the words will bear me out) a moral obligation, which is not felt as such only so far as it is too pleasant.

It is a moral duty for the artist or the inquirer to lead the life of one, and a moral offence when he fails to do so. But on the other hand it is impossible, without violent straining of the facts, to turn these virtues into social virtues or duties to my neighbour. No doubt such virtues do as a rule lead indirectly to the welfare of others, but this is not enough to make them social; their social bearing is indirect, and does not lie in their very essence. The end they aim at is a single end of their own, the content of which does not necessarily involve the good of other men. This we can see from supposing the opposite. If that were true, then it would not be the duty of the inquirer, as such, simply to inquire, or of the artist, as such, simply to produce the best work of art; but each would have to consider ends falling outside his science or art, and would have no right to treat these latter as ends in themselves. 'Nor has he', may be the confident answer. I reply that to me this is a question of fact, and to me it is a fact that the moral consciousness recognizes the perfecting of my intellectual or artistic nature by the production of the proper results, as an end in itself and not merely as a means. The pursuit of these ends, apart from what they lead to, is approved as morally desirable, not perhaps by the theory, but, I think, by the instinctive judgement of all persons worth considering; and if, and while, this fact stands, for me at least it is not affected by doctrines which require that it should be otherwise. To say, without society science and art could not have arisen, is true. To say, apart from society the life of an artist or man of science can not be carried on, is also true; but neither truth goes to show that society is the ultimate end, unless by an argument which takes the basis of a result as its final cause, and which would prove the physical and physiological conditions of society to be the end for which it existed. Man is not man at all unless social, but man is not much above the beasts unless more than social.

If it be said that, morally considered, the realization of the social self is an end, and that of the nonsocial nothing but an outward means, and that hence science and art are not to be pursued independently, no doubt it would be possible to meet such an assertion by argument from and upon its own ground. We might urge that science is most useful, when treated as more than useful. But we decline by doing this to degrade and obscure the question. We repeat that the assertion is both unproven and false, and the decision is left to the moral consciousness of the reader.

And if again it be said that the social self is the one end, but yet none the less science and art are ends in themselves, and to be pursued independently; they are included in the social self, and therefore, as elements in the end, are themselves ends and not mere means—then, in answer, I will not reply that this is false (for indeed I hope it may be true), but only that it is utterly unproven. It is on the assertor that the burden of proof must lie. To us it seems plain that the content of the theoretical self does not in its essence involve relation to others: nothing is easier than to suppose a life of art or speculation which, as far as we can see, though true to itself, has, so far as others are concerned, been sheer waste or even loss, and which knew that it was so. This is a fairly supposable case, and no one, I think, can refuse to enter on it. Was the life immoral? I say, No, it was not *therefore* immoral, but may have been *therefore* moral past ordinary morality. And if I am told Yes, it was moral, but it was social; it did in its essence involve relation to others, because there is a *necessary* connexion (nothing short of this proves the conclusion) between theoretic realization in this and that man, and the realization of him therein and thereby in relation to others, and perhaps also of society as a whole—then I answer, You are asserting in the teeth of appearances; you must prove this necessary connexion, and, I think I may add, you can not do it. What you say may be true, but science, or at all events your science, can not guarantee it; and it is not a truth for the moral consciousness, but leads us further into another region.

Collisions of Moral Elements

Our result at present is as follows. Morality is coextensive with self-realization, as the affirmation of the self which is one with the ideal; and the content of this self is furnished (1) by the objective world of my station and its duties, (2) by the ideal of social, and (3) of nonsocial perfection. And now we have to do with the question, How do these spheres stand to one another? And this is in some ways an awkward question, because it brings up practical everyday difficulties. They are something of this sort. May a man, for the sake of science or art, venture on acts of commission or omission which in any one else would be immoral; or, to put it coarsely, may he be what is generally called a bad man, may he trample on ordinary morality, in order that he may be a good artist? Or again, if the perhaps less familiar question of the relation of (1) to (2) comes up, the doubt is, Must I do the work that lies next me in the world, and so serve society, even, as it seems, to the detriment of my own moral being? May I adopt a profession considered moral by the world, but which, judged by my ideal, can not be called moral?

The first point to which we must call attention is that all these are cases of colliding duties. In none of them is there a contest between the claims of morality and of something else not morality. In the moral sphere such a contest is impossible and meaningless. We have in all of them a conflict between moral duties which are taken to exclude one another, e.g. my moral duty as artist on the one hand and as father of a family on the other, and so on: we have nothing to do with examples where morality is neglected or opposed in the name of anything else than an other and higher morality.

And the second point, which has engaged us before (pp. 156–59, 193 foll. [pp. 62–64 and 65 foll. in this volume]), and on which we desire to insist with emphasis, is, that cases of collision of duties are not scientific but practical questions. Moral science has nothing whatever to do with the settlement of them; that would belong, did such a thing exist, to the moral art. The difficulties of collisions are not scientific problems; they arise from the complexity of individual cases, and this can be dealt with solely by practical insight, not by abstract conceptions and discursive reasoning. It is no use knowing that one class of duties is in the abstract higher than another: moral practice is not *in abstracto*, and the highest moral duty for *me* is *my* duty; *my* duty being the one which lies next me, and perhaps not the one which would be the highest, supposing it were mine. The man who can give moral advice is the man of experience, who, from his own knowledge and by sympathy, can transport himself into another's case; who knows the heart and sees through moral illusion; and the man of mere theory is in the practical sphere a useless and dangerous pedant.

And now in particular the relation of the two ideal spheres to the real sphere is precisely what subsists inside the real sphere between its own elements. We saw (pp. 156–57 [pp. 62–64 in this volume]) that, as in no one action can all duties be fulfilled, in every action some duties must be neglected. The question is what duty is to be done and left undone *here;* and so in the world of my station neglect of duties is allowed. And, apart from the difficulty (often the impossibility) of distinguishing omission and commission from a moral point of view, we saw (ibid.) that positive breaches of moral law were occasionally moral. And hence if an artist or man of science considers himself called upon, by his duty to art or science, to neglect, or to commit a breach of, ordinary morality, we must say that, in the abstract and by itself, that is not to be condemned. It is a case of colliding duties, such as happens every day in other fields, and its character is not different because extraordinary.

And further, if a claim be set up, on the ground of devotion to no common end, to be judged in one's life by no common standard, we must admit that already within the sphere of my station that claim is usually allowed. We excuse in a soldier or sailor what we do not excuse in others, from whom the same duties are not expected. The morality of the pushing man of business, and still more of the lawyer and the diplomatist in the exercise of their calling, is not measured by the standard of common life; and so, when the service of the ideal is appealed to in justification of neglect and breaches of law, we say that the claim is valid in itself, the abstract right is undeniable, the case is a case of collision, and the question of moral justification is a question of particular fact.

Collision of duties carries all this with it on the one side, but we must not forget what it carries on the other. In raising that excuse we are saying, 'I neglect duty because of duty'; and this means we recognize two duties, one higher than the other. And first it implies that we are acting, not to please ourselves, but because we are bound by what we consider moral duty. It implies again that we consider what we break through or pass by, not as a trifle, but as a serious moral claim, which we disregard solely because, if we do not do so, it prevents us from performing our superior service.

Common social morality is the basis of human life. It is specialized in particular functions of society, and upon its foundation are erected the ideals of a higher social perfection and of the theoretic life; but common morality remains both the cradle and protecting nurse of its aspiring offspring, and, if we ever forget that, we lie open to the charge of ingratitude and baseness. Some neglect is unavoidable; but open and direct outrage on the standing moral institutions which make society and human life what it is, can be justified (I do not say condoned) only on the plea of overpowering moral necessity. And the individual should remember that the will for good, if weakened in one place, runs the greatest risk of being weakened in all.

Our result then is that ideal morality stands on the basis of [the] social, that its relation thereto is the same relation that subsists within the social sphere, and that everywhere, since duty has to give way to duty, neglect and breaches of ordinary in the name of higher morality are justifiable in the abstract (and that is all we are concerned with); but if the claim be set up, on account of devotion to the ideal, for liberty to act thus not in the name of moral necessity, or to forget that what we break through or disregard is in itself to be respected, such a claim is without the smallest moral justification.[23]

The highest type we can imagine is the man who, on the basis of everyday morality, aims at the ideal perfection of it, and on this double basis strives to realize a nonsocial ideal. But where collisions arise, there, we must repeat, it is impossible for mere theory to offer a solution, not only because the perception which decides is not a mere intellectual perception, but because no general solution of individual difficulties is possible.

Notes

1. Bradley has just completed his critique of "Kantian" formalism.— C.W.G.

2. [Collision of Duties: cf. p. 226, line 13 (p. 78 in this volume).]

3. [Except of course the universal law to do the best we can in the circumstances.]

4. The story of the imprisoned Italian who, knowing that he was being drugged to disorder his intellect and cause him to betray his comrades, opened a vein, is a good instance. It is a duty for various persons continually to give themselves to certain or well-nigh certain death, and no one has ever called it anything but heroically right and dutiful. Excusable

killing is illustrated by the well-known story told in the Indian Mutiny of the husband who killed his wife. Rebellions and mutinies need no illustration. It is noticeable that Berkeley urged passive obedience on the ground that a moral law was absolute. [Cf. Berkeley, *Passive Obedience* (Fraser's edition of Berkeley's Works, volume iii).]

5. We shall come upon this again in Essays V and VI. [The relevant sections are included in this chapter.—C.W.G.]

6. I must ask the reader here not to think of 'Intuitionalism', or of 'Organs of the Absolute', or of anything else of the sort. 'Intuitive' is used here as the opposite of 'reflective' or 'discursive', 'intuition' as the opposite of 'reasoning' or 'explicit inferring'. If the reader dislike the word, he may substitute 'perception' or 'sense', if he will; but then he must remember that neither are to exclude the intellectual, the understanding and its implicit judgements and inferences.

7. It is right to remark that second thoughts are often the offspring of wrong desire, but not always so. They may arise from collisions, and in these cases we see how little is to be done by theoretical deduction.

8. Not, perhaps, on *all* matters. Nor, again, will it do to say that *everywhere* women are pre-eminently intuitive, and men discursive. But in *practical* matters there seems not much doubt that it is so.

9. Every act has, of course, many sides, many relations, many 'points of view from which it may be regarded', and so many qualities. There are always several principles under which you can bring it, and hence there is not the smallest difficulty in exhibiting it as the realization of either right or wrong. No act in the world is without *some* side capable of being subsumed under a good rule; e.g. theft is economy, care for one's relations, protest against bad institutions, really doing oneself but justice, etc.; and, if all else fails, it probably saves us from something worse, and therefore is good. Cowardice is prudence and a duty, courage rashness and a vice, and so on. The casuist must have little ingenuity, if there is anything he fails to justify or condemn according to his order. And the vice of casuistry is that, attempting to decide the particulars of morality by the deductions of the reflective understanding, it at once degenerates into finding a good reason for what you mean to do. You have principles of all sorts, and the case has all sorts of sides; *which* side is the essential side, and which principle is *the* principle *here,* rests in the end on your mere private choice; and that is determined by heaven knows what. No *reasoning* will tell you which the moral point of view *here* is. Hence the necessary immorality and the ruinous effects of practical casuistry. (Casuistry used not as a guide to conduct, but as a means to the theoretical investigation of moral principles, the casuistry used to discover the principle *from* the fact, and not to deduce the fact from the principle—is, of course, quite another thing.) Our moralists do not like casuistry; but if the current notion that moral philosophy has to tell you what to do is well founded, then casuistry, so far as I can see, at once follows, or should follow.

But the ordinary moral judgement is not discursive. It does not look to the right and left, and, considering the case from all its sides, consciously subsume under one principle. When the case is presented, it fixes on one quality in the act, referring that unconsciously to one principle, in which it feels the whole of itself, and sees that whole in a single side of the act. So far as right and wrong are concerned, it can perceive nothing but *this* quality of *this* case, and anything else it refuses to try to perceive. Practical morality means singlemindedness, the having one idea; it means what in other spheres [not wholly so, for in the intellectual world also the relevant, and here essential, is often seized intuitively and not reflectively] would be the greatest narrowness. Point out to a man of simple morals that the case has other sides than the one he instinctively fixes on, and he suspects you wish to corrupt him. And so you probably would if you went on. Apart from bad example, the readiest way to debauch the morality of any one is, on the side of principle, to confuse them by forcing them to see in all moral and immoral acts other sides and points of view, which alter the character of each; and, on the side of particulars, to warp their instinctive apprehension through personal affection for yourself or some other individual.

10. It is worth while in this connexion to refer to the custom some persons have (and find useful) of calling before the mind, when in doubt, a known person of high character and quick judgement, and thinking what they would have done. This no doubt both delivers the mind from private considerations and also is to act in the spirit of the other person (so far as we know it), i.e. from the general basis of his acts (certainly *not* the mere memory of his particular acts, or such memory plus inference).

11. One of these would be as to how progress in morality is made.

12. [This is too optimistic.]

13. ["Society," see pp. 173–74, 222–23 (pp. 75–76 in this volume).]

14. I may remark on this (after Erdmann, and I suppose Plato) that collisions of duties are avoided mostly by each man keeping to his own immediate duties, and not trying to see from the point of view of other stations than his own.

15. [Cf. p. 228, and note, pp. 244, 218–19 (p. 72 in this volume).]

16. Expressed in other language the objection is, 'There is a sphere of rights which falls outside the sphere of duty, or else it will follow that all my rights are my duties, which is absurd'.—For the answer, see p. 210 [in *Ethical Studies*]. Here we may say, it is right and a duty that the sphere of indifferent detail should exist. It is a duty that I should develop my nature by private choice therein. Therefore, *because* that is a duty, it is a duty *not* to make a duty of every detail; and thus in every detail I have done my duty.

17. It may even be my moral duty to be religious in the sense of acting with a view to the support and maintenance of the religious consciousness, the faith which is to reissue in religious-moral practice. Hence though morality, as we shall see, does not include everything, yet nothing in another sense falls outside of it.

18. On the genesis of the ideal self and of the good self, or the self

whose will is identified with its ideal, we shall say what seems necessary in other connexions.

19. I may remark that a duty which is *not* a duty to myself can not possibly be a moral duty. When we hear of self-regarding duties we should ask what is meant. [This *might* mean a duty towards myself as this or that member of society.] A '*self*-regarding duty' in one sense of the word says no more than 'a duty'; in another sense it says 'a duty which is the direct opposite of what a duty is', i.e. a *selfish* duty: or again, it means a nonsocial duty. Confusion on this head leads to serious mistakes.

20. Virtues such as chastity, which might be practised in solitude, are either negative of the bad self, or conditions of the good will. If you wrongly consider them by themselves, they are not positively desirable. We may call them, if we will, the 'ascetic virtues'.

21. The difficulty everywhere is, Is the embodiment used to fire the imagination, while the type is not that of this or that individual; or is it otherwise? The solution is to be found in the answer to the question, Is the impersonation modified; and if modified, how, and by what, and to suit what is it modified?

22. Morality, on its own ground at least, knows nothing of a universal and invisible self, in which all members are real, which they realize in their own gifts and graces, and in realizing which they realize the other members. Humanity as an organic whole, if a possible point of view, is not strictly speaking a moral point of view.

23. I have not entered on the questions whether as a fact breaches of common morality are demanded by the service of the ideal, and, if so, when they are to be committed. The first is a matter of fact it would not profit us to discuss in connexion with the abstract question; and the second in our opinion can not be theoretically determined. Which duty or duties weigh heaviest in this or that case is an affair for perception, not reasoning. We may remark, however, that the doctrine of the text will not be found to err on the side of laxity.

4

Prima Facie *Duties*

Sir David Ross

The real point at issue between hedonism and utilitarianism on the one hand and their opponents on the other is not whether 'right' means 'productive of so and so'; for it cannot with any plausibility be maintained that it does. The point at issue is that to which we now pass, viz. whether there is any general character which makes right acts right, and if so, what it is. Among the main historical attempts to state a single characteristic of all right actions which is the foundation of their rightness are those made by egoism and utilitarianism. But I do not propose to discuss these, not because the subject is unimportant, but because it has been dealt with so often and so well already, and because there has come to be so much agreement among moral philosophers that neither of these theories is satisfactory. A much more attractive theory has been put forward by Professor Moore[1]: that what makes actions right is that they are productive of more *good* than could have been produced by any other action open to the agent.[2]

Reprinted from *The Right and the Good* by W D Ross (1930) by permission of Oxford University Press. This chapter is from pp. 16–34 and 41–42 of *The Right and the Good*. The chapter title is provided by C.W.G. Page references in the notes are to *The Right and the Good*.

This theory is in fact the culmination of all the attempts to base rightness on productivity of some sort of result. The first form this attempt takes is the attempt to base rightness on conduciveness to the advantage or pleasure of the agent. This theory comes to grief over the fact, which stares us in the face, that a great part of duty consists in an observance of the rights and a furtherance of the interests of others, whatever the cost to ourselves may be. Plato and others may be right in holding that a regard for the rights of others never in the long run involves a loss of happiness for the agent, that 'the just life profits a man'. But this, even if true, is irrelevant to the rightness of the act. As soon as a man does an action *because* he thinks he will promote his own interests thereby, he is acting not from a sense of its rightness but from self-interest.

To the egoistic theory hedonistic utilitarianism supplies a much-needed amendment. It points out correctly that the fact that a certain pleasure will be enjoyed by the agent is no reason why he *ought* to bring it into being rather than an equal or greater pleasure to be enjoyed by another, though, human nature being what it is, it makes it not unlikely that he *will* try to bring it into being. But hedonistic utilitarianism in its turn needs a correction. On reflection it seems clear that pleasure is not the only thing in life that we think good in itself, that for instance we think the possession of a good character, or an intelligent understanding of the world, as good or better. A great advance is made by the substitution of 'productive of the greatest good' for 'productive of the greatest pleasure'.

Not only is this theory more attractive than hedonistic utilitarianism, but its logical relation to that theory is such that the latter could not be true unless *it* were true, while it might be true though hedonistic utilitarianism were not. It is in fact one of the logical bases of hedonistic utilitarianism. For the view that what produces the maximum pleasure is right has for its bases the views (1) that what produces the maximum good is right, and (2) that pleasure is the only thing good in itself. If they were not assuming that what produces the maximum *good* is right, the utilitarians' attempt to show that pleasure is the only thing good in itself, which is in fact the point they take most pains to establish, would have been quite irrelevant to their attempt to prove that only what produces the maximum *pleasure* is right. If, therefore, it can be shown that productivity of the maximum good is not what makes all right actions right, we shall *a fortiori* have refuted hedonistic utilitarianism.

When a plain man fulfils a promise because he thinks he ought to do so, it seems clear that he does so with no thought of its total consequences, still less with any opinion that these are likely to be the best possible. He thinks in fact much more of the past than of the future. What makes him think it right to act in a certain way is the fact that he has promised to do so—that and, usually, nothing more. That his act will produce the best possible consequences is not his reason for calling it right. What lends colour to the theory we are examining, then, is not the actions (which form probably a great majority of our actions) in which some such reflection as 'I have promised' is the only reason we give ourselves for thinking a certain action right, but the exceptional cases in which the consequences of fulfilling a promise (for instance) would be so disastrous to others that we judge it right not to do so. It must of course be admitted that such cases exist. If I have promised to meet a friend at a particular time for some trivial purpose, I should certainly think myself justified in breaking my engagement if by doing so I could prevent a serious accident or bring relief to the victims of one. And the supporters of the view we are examining hold that my thinking so is due to my thinking that I shall bring more good into existence by the one action than by the other. A different account may, however, be given of the matter, an account which will, I believe, show itself to be the true one. It may be said that besides the duty of fulfilling promises I have and recognize a duty of relieving distress,[3] and that when I think it right to do the latter at the cost of not doing the former, it is not because I think I shall produce more good thereby but because I think it the duty which is in the circumstances more of a duty. This account surely corresponds much more closely with what we really think in such a situation. If, so far as I can see, I could bring equal amounts of good into being by fulfilling my promise and by helping some one to whom I had made no promise, I should not hesitate to regard the former as my duty. Yet on the view that what is right is right because it is productive of the most good I should not so regard it.

There are two theories, each in its way simple, that offer a solution of such cases of conscience. One is the view of Kant, that there are certain duties of perfect obligation, such as those of fulfilling promises, of paying debts, of telling the truth, which admit of no exception whatever in favour of duties of imperfect obligation, such as that of relieving distress. The other is the view of, for instance,

Professor Moore and Dr. Rashdall,[4] that there is only the duty of producing good, and that all 'conflicts of duties' should be resolved by asking 'by which action will most good be produced?' But it is more important that our theory fit the facts than that it be simple, and the account we have given above corresponds (it seems to me) better than either of the simpler theories with what we really think, viz. that normally promise-keeping, for example, should come before benevolence, but that when and only when the good to be produced by the benevolent act is very great and the promise comparatively trivial, the act of benevolence becomes our duty.

In fact the theory of 'ideal utilitarianism', if I may for brevity refer so to the theory of Professor Moore, seems to simplify unduly our relations to our fellows. It says, in effect, that the only morally significant relation in which my neighbours stand to me is that of being possible beneficiaries by my action.[5] They do stand in this relation to me, and this relation is morally significant. But they may also stand to me in the relation of promisee to promiser, of creditor to debtor, of wife to husband, of child to parent, of friend to friend, of fellow countryman to fellow countryman, and the like; and each of these relations is the foundation of a *prima facie* duty, which is more or less incumbent on me according to the circumstances of the case. When I am in a situation, as perhaps I always am, in which more than one of these *prima facie* duties is incumbent on me, what I have to do is to study the situation as fully as I can until I form the considered opinion (it is never more) that in the circumstances one of them is more incumbent than any other; then I am bound to think that to do this *prima facie* duty is my duty *sans phrase* in the situation.

I suggest '*prima facie* duty' or 'conditional duty' as a brief way of referring to the characteristic (quite distinct from that of being a duty proper) which an act has, in virtue of being of a certain kind (e.g. the keeping of a promise), of being an act which would be a duty proper if it were not at the same time of another kind which is morally significant. Whether an act is a duty proper or actual duty depends on *all* the morally significant kinds it is an instance of. The phrase '*prima facie* duty' must be apologized for, since (1) it suggests that what we are speaking of is a certain kind of duty, whereas it is in fact not a duty, but something related in a special way to duty. Strictly speaking, we want not a phrase in which duty is qualified by an adjective, but a separate noun. (2) '*Prima' facie* suggests that one is speaking only of an appearance which a moral situation presents at first sight, and which may turn out to be illusory; whereas what I am speaking of is an objective fact involved in the nature of the situation, or more strictly in an element of its nature, though not, as duty proper does, arising from its *whole* nature. I can, however, think of no term which fully meets the case. 'Claim' has been suggested by Professor Prichard.[6] The word 'claim' has the advantage of being quite a familiar one in this connexion, and it seems to cover much of the ground. It would be quite natural to say, 'a person to whom I have made a promise has a claim on me', and also, 'a person whose distress I could relieve (at the cost of breaking the promise) has a claim on me'. But (1) while 'claim' is appropriate from *their* point of view, we want a word to express the corresponding fact from the agent's point of view—the fact of his being subject to claims that can be made against him; and ordinary language provides us with no such correlative to 'claim'. And (2) (what is more important) 'claim' seems inevitably to suggest two persons, one of whom might make a claim on the other; and while this covers the ground of social duty, it is inappropriate in the case of that important part of duty which is the duty of cultivating a certain kind of character in oneself. It would be artificial, I think, and at any rate metaphorical, to say that one's character has a claim on oneself.

There is nothing arbitrary about these *prima facie* duties. Each rests on a definite circumstance which cannot seriously be held to be without moral significance. Of *prima facie* duties I suggest, without claiming completeness or finality for it, the following division.[7]

(1) Some duties rest on previous acts of my own. These duties seem to include two kinds, (*a*) those resting on a promise or what may fairly be called an implicit promise, such as the implicit undertaking not to tell lies which seems to be implied in the act of entering into conversation (at any rate by civilized men), or of writing books that purport to be history and not fiction. These may be called the duties of fidelity. (*b*) Those resting on a previous wrongful act. These may be called the duties of reparation. (2) Some rest on previous acts of other men, i.e. services done by them to me. These may be loosely described as the duties of gratitude.[8] (3) Some rest on the fact or possibility of a distribution of pleasure or happiness (or of the means thereto) which is not in accordance with the merit of the persons concerned; in such cases there arises a duty to upset or prevent such a distribution. These are the duties of justice. (4) Some rest on the mere fact that there are other beings in the world whose condition we can make better in respect of virtue, or of intel-

ligence, or of pleasure. These are the duties of beneficence. (5) Some
rest on the fact that we can improve our own condition in respect
of virtue or of intelligence. These are the duties of self-improvement. (6) I think we should distinguish from (4) the duties that may
be summed up under the title of 'not injuring others'. No doubt to
injure others is incidentally to fail to do them good; but it seems to
me clear that nonmaleficence is apprehended as a duty distinct from
that of beneficence, and as a duty of a more stringent character. It
will be noticed that this alone among the types of duty has been
stated in a negative way. An attempt might no doubt be made to
state this duty, like the others, in a positive way. It might be said
that it is really the duty to prevent ourselves from acting either from
an inclination to harm others or from an inclination to seek our own
pleasure, in doing which we should incidentally harm them. But on
reflection it seems clear that the primary duty here is the duty not
to harm others, this being a duty whether or not we have an inclination that if followed would lead to our harming them; and that
when we have such an inclination the primary duty not to harm
others gives rise to a consequential duty to resist the inclination.
The recognition of this duty of nonmaleficence is the first step on
the way to the recognition of the duty of beneficence; and that
accounts for the prominence of the commands 'thou shalt not kill',
'thou shalt not commit adultery', 'thou shalt not steal', 'thou shalt
not bear false witness', in so early a code as the Decalogue. But even
when we have come to recognize the duty of beneficence, it appears
to me that the duty of nonmaleficence is recognized as a distinct
one, and as *prima facie* more binding. We should not in general consider it justifiable to kill one person in order to keep another alive,
or to steal from one in order to give alms to another.

   The essential defect of the 'ideal utilitarian' theory is that it
ignores, or at least does not do full justice to, the highly personal
character of duty. If the only duty is to produce the maximum of
good, the question who is to have the good—whether it is myself,
or my benefactor, or a person to whom I have made a promise to
confer that good on him, or a mere fellow man to whom I stand in
no such special relation—should make no difference to my having
a duty to produce that good. But we are all in fact sure that it makes
a vast difference.

   One or two other comments must be made on this provisional
list of the divisions of duty. (1) The nomenclature is not strictly correct. For by 'fidelity' or 'gratitude' we mean, strictly, certain states

of motivation; and, as I have urged, it is not our duty to have certain motives, but to do certain acts. By 'fidelity', for instance, is meant, strictly, the disposition to fulfil promises and implicit promises *because we have made them.* We have no general word to cover the actual fulfilment of promises and implicit promises *irrespective of motive;* and I use 'fidelity', loosely but perhaps conveniently, to fill this gap. So too I use 'gratitude' for the returning of services, irrespective of motive. The term 'justice' is not so much confined, in ordinary usage, to a certain state of motivation, for we should often talk of a man as acting justly even when we did not think his motive was the wish to do what was just simply for the sake of doing so. Less apology is therefore needed for our use of 'justice' in this sense. And I have used the word 'beneficence' rather than 'benevolence', in order to emphasize the fact that it is our duty to do certain things, and not to do them from certain motives.

(2) If the objection be made, that this catalogue of the main types of duty is an unsystematic one resting on no logical principle, it may be replied, first, that it makes no claim to being ultimate. It is a *prima facie* classification of the duties which reflection on our moral convictions seems actually to reveal. And if these convictions are, as I would claim that they are, of the nature of knowledge, and if I have not misstated them, the list will be a list of authentic conditional duties, correct as far as it goes though not necessarily complete. The list of *goods* put forward by the rival theory is reached by exactly the same method—the only sound one in the circumstances—viz. that of direct reflection on what we really think. Loyalty to the facts is worth more than a symmetrical architectonic or a hastily reached simplicity. If further reflection discovers a perfect logical basis for this or for a better classification, so much the better.

(3) It may, again, be objected that our theory that there are these various and often conflicting types of *prima facie* duty leaves us with no principle upon which to discern what is our actual duty in particular circumstances. But this objection is not one which the rival theory is in a position to bring forward. For when we have to choose between the production of two heterogeneous goods, say knowledge and pleasure, the 'ideal utilitarian' theory can only fall back on an opinion, for which no logical basis can be offered, that one of the goods is the greater; and this is no better than a similar opinion that one of two duties is the more urgent. And again, when we consider the infinite variety of the effects of our actions in the way of pleasure, it must surely be admitted that the claim which

hedonism sometimes makes, that it offers a readily applicable criterion of right conduct, is quite illusory.

I am unwilling, however, to content myself with an *argumentum ad hominem,* and I would contend that in principle there is no reason to anticipate that every act that is our duty is so for one and the same reason. Why should two sets of circumstances, or one set of circumstances, *not* possess different characteristics, any one of which makes a certain act our *prima facie* duty? When I ask what it is that makes me in certain cases sure that I have a *prima facie* duty to do so and so, I find that it lies in the fact that I have made a promise; when I ask the same question in another case, I find the answer lies in the fact that I have done a wrong. And if on reflection I find (as I think I do) that neither of these reasons is reducible to the other, I must not on any a priori ground assume that such a reduction is possible.

An attempt may be made to arrange in a more systematic way the main types of duty which we have indicated. In the first place it seems self-evident that if there are things that are intrinsically good, it is *prima facie* a duty to bring them into existence rather than not to do so, and to bring as much of them into existence as possible. It will be argued in our fifth chapter [of *The Right and the Good*] that there are three main things that are intrinsically good—virtue, knowledge, and, with certain limitations, pleasure. And since a given virtuous disposition, for instance, is equally good whether it is realized in myself or in another, it seems to be my duty to bring it into existence whether in myself or in another. So too with a given piece of knowledge.

The case of pleasure is difficult; for while we clearly recognize a duty to produce pleasure for others, it is by no means so clear that we recognize a duty to produce pleasure for ourselves. This appears to arise from the following facts. The thought of an act as our duty is one that presupposes a certain amount of reflection about the act; and for that reason does not normally arise in connexion with acts towards which we are already impelled by another strong impulse. So far, the cause of our not thinking of the promotion of our own pleasure as a duty is analogous to the cause which usually prevents a highly sympathetic person from thinking of the promotion of the pleasure of others as a duty. He is impelled so strongly by direct interest in the well-being of others towards promoting their pleasure that he does not stop to ask whether it is his duty to promote it; and we are all impelled so strongly towards the promotion of our own pleasure that we do not stop to ask whether it is a duty or not. But there is a further reason why even when we stop to think about the matter it does not usually present itself as a duty: viz. that, since the performance of most of our duties involves the giving up of some pleasure that we desire, the doing of duty and the getting of pleasure for ourselves come by a natural association of ideas to be thought of as incompatible things. This association of ideas is in the main salutary in its operation, since it puts a check on what but for it would be much too strong, the tendency to pursue one's own plea-sure without thought of other considerations. Yet if pleasure is good, it seems in the long run clear that it is right to get it for our-selves as well as to produce it for others, when this does not involve the failure to discharge some more stringent *prima facie* duty. The question is a very difficult one, but it seems that this conclusion can be denied only on one or other of three grounds: (1) that pleasure is not *prima facie* good (i.e. good when it is neither the actualization of a bad disposition nor undeserved), (2) that there is no *prima facie* duty to produce as much that is good as we can, or (3) that though there is a *prima facie* duty to produce other things that are good, there is no *prima facie* duty to produce pleasure which will be enjoyed by ourselves. I give reasons later[9] for not accepting the first contention. The second hardly admits of argument but seems to me plainly false. The third seems plausible only if we hold that an act that is pleasant or brings pleasure to ourselves must for that reason not be a duty; and this would lead to paradoxical consequences, such as that if a man enjoys giving pleasure to others or working for their moral improvement, it cannot be his duty to do so. Yet it seems to be a very stubborn fact, that in our ordinary consciousness we are not aware of a duty to get pleasure for ourselves; and by way of partial explanation of this I may add that though, as I think, one's own pleasure is a good and there is a duty to produce it, it is only if we *think* of our own pleasure not as simply our own pleasure, but as an objective good, something that an impartial spectator would approve, that we can think of the getting it as a duty; and we do not habitually think of it in this way.

If these contentions are right, what we have called the duty of beneficence and the duty of self-improvement rest on the same ground. No different principles of duty are involved in the two cases. If we feel a special responsibility for improving our own char-acter rather than that of others, it is not because a special principle is involved, but because we are aware that the one is more under our control than the other. It was on this ground that Kant expressed the practical law of duty in the form 'seek to make yourself good and other people happy'. He was so persuaded of the internality of virtue that he regarded any attempt by one person to produce virtue in another as bound to produce, at most, only a counterfeit of virtue, the doing of externally right acts not from the true principle of virtuous action but out of regard to another person. It must be admitted that one man cannot compel another to be virtuous; compulsory virtue would just not be virtue. But experience clearly shows that Kant overshoots the mark when he contends that one man cannot do anything to *promote* virtue in another, to bring such influences to bear upon him that his own response to them is more likely to be virtuous than his response to other influences would have been. And our duty to do this is not different in kind from our duty to improve our own characters.

It is equally clear, and clear at an earlier stage of moral development, that if there are things that are bad in themselves we ought, *prima facie,* not to bring them upon others; and on this fact rests the duty of nonmaleficence.

The duty of justice is particularly complicated, and the word is used to cover things which are really very different—things such as the payment of debts, the reparation of injuries done by oneself to another, and the bringing about of a distribution of happiness between other people in proportion to merit. I use the word to denote only the last of these three. In the fifth chapter [of *The Right and the Good*] I shall try to show that besides the three (comparatively) simple goods, virtue, knowledge, and pleasure, there is a more complex good, not reducible to these, consisting in the proportionment of happiness to virtue. The bringing of this about is a duty which we owe to all men alike, though it may be reinforced by special responsibilities that we have undertaken to particular men. This, therefore, with beneficence and self-improvement, comes under the general principle that we should produce as much good as possible, though the good here involved is different in kind from any other.

But besides this general obligation, there are special obligations. These may arise, in the first place, incidentally, from acts which were not essentially meant to create such an obligation, but which nevertheless create it. From the nature of the case such acts may be of two kinds—the infliction of injuries on others, and the acceptance of benefits from them. It seems clear that these put us under a special obligation to other men, and that only these acts can do so incidentally. From these arise the twin duties of reparation and gratitude.

And finally there are special obligations arising from acts the very intention of which, when they were done, was to put us under such an obligation. The name for such acts is 'promises'; the name is wide enough if we are willing to include under it implicit promises, i.e. modes of behaviour in which without explicit verbal promise we intentionally create an expectation that we can be counted on to behave in a certain way in the interest of another person.

These seem to be, in principle, all the ways in which *prima facie* duties arise. In actual experience they are compounded together in highly complex ways. Thus, for example, the duty of obeying the laws of one's country arises partly (as Socrates contends in the *Crito*) from the duty of gratitude for the benefits one has received from it; partly from the implicit promise to obey which seems to be involved in permanent residence in a country whose laws we know we are *expected* to obey, and still more clearly involved when we ourselves invoke the protection of its laws (this is the truth underlying the doctrine of the social contract); and partly (if we are fortunate in our country) from the fact that its laws are potent instruments for the general good.

Or again, the sense of a general obligation to bring about (so far as we can) a just apportionment of happiness to merit is often greatly reinforced by the fact that many of the existing injustices are due to a social and economic system which we have, not indeed created, but taken part in and assented to; the duty of justice is then reinforced by the duty of reparation.

It is necessary to say something by way of clearing up the relation between *prima facie* duties and the actual or absolute duty to do one particular act in particular circumstances. If, as almost all moralists except Kant are agreed, and as most plain men think, it is sometimes right to tell a lie or to break a promise, it must be maintained that there is a difference between *prima facie* duty and actual or absolute duty. When we think ourselves justified in breaking, and indeed morally obliged to break, a promise in order to relieve some one's distress, we do not for a moment cease to recognize a *prima facie* duty to keep our promise, and this leads us to feel, not indeed shame or repentance, but certainly compunction, for behaving as we do; we recognize, further, that it is our duty to make up somehow to the promisee for the breaking of the promise. We have to distinguish from the characteristic of being our duty that of tending to be our duty. Any act that we do contains various elements in virtue of which it falls under various categories. In virtue of being the breaking of a promise, for instance, it tends to be wrong; in virtue of being an instance of relieving distress it tends to be right. Tendency to be one's duty may be called a parti-resultant attribute, i.e. one which belongs to an act in virtue of some one component in its nature. *Being* one's duty is a toti-resultant attribute, one which belongs to an act in virtue of its whole nature and of nothing less than this.[10] This distinction between parti-resultant and toti-resultant attributes is one which we shall meet in another context also.[11]

Another instance of the same distinction may be found in the operation of natural laws. Qua subject to the force of gravitation towards some other body, each body tends to move in a particular direction with a particular velocity; but its actual movement depends on *all* the forces to which it is subject. It is only by recognizing this distinction that we can preserve the absoluteness of laws of nature, and only by recognizing a corresponding distinction that we can preserve the absoluteness of the general principles of morality. But an important difference between the two cases must be pointed out. When we say that in virtue of gravitation a body tends to move in a certain way, we are referring to a causal influence actually exercised on it by another body or other bodies. When we say that in virtue of being deliberately untrue a certain remark tends to be wrong, we are referring to no causal relation, to no relation that involves succession in time, but to such a relation as connects the various attributes of a mathematical figure. And if the word 'tendency' is thought to suggest too much a causal relation, it is better to talk of certain types of act as being *prima facie* right or wrong (or of different persons as having different and possibly conflicting claims upon us), than of their tending to be right or wrong.

Something should be said of the relation between our apprehension of the *prima facie* rightness of certain types of act and our mental attitude towards particular acts. It is proper to use the word 'apprehension' in the former case and not in the latter. That an act, qua fulfilling a promise, or qua effecting a just distribution of good, or qua returning services rendered, or qua promoting the good of others, or qua promoting the virtue or insight of the agent, is *prima facie* right, is self-evident; not in the sense that it is evident from the beginning of our lives, or as soon as we attend to the proposition for the first time, but in the sense that when we have reached sufficient mental maturity and have given sufficient attention to the proposition it is evident without any need of proof, or of evidence beyond itself. It is self-evident just as a mathematical axiom, or the validity of a form of inference, is evident. The moral order expressed in these propositions is just as much part of the fundamental nature of the universe (and, we may add, of any possible universe in which there were moral agents at all) as is the spatial or numerical structure expressed in the axioms of geometry or arithmetic. In our confidence that these propositions are true there is involved the same trust in our reason that is involved in our confidence in mathematics; and we should have no justification for trusting it in the latter sphere and distrusting it in the former. In both cases we are dealing with propositions that cannot be proved, but that just as certainly need no proof.

Some of these general principles of *prima facie* duty may appear to be open to criticism. It may be thought, for example, that the principle of returning good for good is a falling off from the Christian principle, generally and rightly recognized as expressing the highest morality, of returning good for evil. To this it may be replied that I do not suggest that there is a principle commanding us to return good for good and forbidding us to return good for evil, and that I do suggest that there is a positive duty to seek the good of all men. What I maintain is that an act in which good is returned for good is recognized as *specially* binding on us just because it is of that character, and that *ceteris paribus* any one would think it his duty to help his benefactors rather than his enemies, if he could not do both; just as it is generally recognized that *ceteris paribus* we should pay our debts rather than give our money in charity, when we cannot do both. A benefactor is not only a man, calling for our effort on his behalf on that ground, but also our benefactor, calling for our *special* effort on *that* ground.

Our judgements about our actual duty in concrete situations have none of the certainty that attaches to our recognition of the general principles of duty. A statement is certain, i.e. is an expression of knowledge, only in one or other of two cases: when it is either self-evident, or a valid conclusion from self-evident premisses. And our judgements about our particular duties have neither of these characters. (1) They are not self-evident. Where a possible act is seen to have two characteristics, in virtue of one of which it is *prima facie* right, and in virtue of the other *prima facie* wrong, we are (I think) well aware that we are not certain whether we ought or ought not to do it; that whether we do it or not, we are taking a moral risk. We come in the long run, after consideration, to think one duty more pressing than the other, but we do not feel certain that it is so. And though we do not always recognize that a possible act has two such characteristics, and though there *may* be cases in which it has not, we are never certain that any particular possible act has not, and therefore never certain that it is right, nor certain that it is wrong. For, to go no further in the analysis, it is enough to point out that any particular act will in all probability in the course of time contribute to the bringing about of good or of evil for many human beings, and thus have a *prima facie* rightness or wrongness of which we know nothing. (2) Again, our judgements about our particular duties are not logical conclusions from self-evident premisses. The only possible premisses would be the general principles stating their *prima facie* rightness or wrongness qua having the different characteristics they do have; and even if we could (as we cannot) apprehend the extent to which an act will tend on the one hand, for example, to bring about advantages for our benefactors, and on the other hand to bring about disadvantages for fellow men who are not our benefactors, there is no principle by which we can draw the conclusion that it is on the whole right or on the whole wrong. In this respect the judgement as to the rightness of a particular act is just like the judgement as to the beauty of a particular natural object or work of art. A poem is, for instance, in respect of certain qualities beautiful and in respect of certain others not beautiful; and our judgement as to the degree of beauty it possesses on the whole is never reached by logical reasoning from the apprehension of its particular beauties or particular defects. Both in this and in the moral case we have more or less probable opinions which are not logically justified conclusions from the general principles that are recognized as self-evident.

There is therefore much truth in the description of the right act as a fortunate act. If we cannot be certain that it is right, it is our good fortune if the act we do is the right act. This consideration does not, however, make the doing of our duty a mere matter of chance. There is a parallel here between the doing of duty and the doing of what will be to our personal advantage. We never *know* what act will in the long run be to our advantage. Yet it is certain that we are more likely in general to secure our advantage if we estimate to the best of our ability the probable tendencies of our actions in this respect, than if we act on caprice. And similarly we are more likely to do our duty if we reflect to the best of our ability on the *prima facie* rightness or wrongness of various possible acts in virtue of the characteristics we perceive them to have, than if we act without reflection. With this greater likelihood we must be content.

Many people would be inclined to say that the right act for me is not that whose general nature I have been describing, viz. that which if I were omniscient I should see to be my duty, but that which on all the evidence available to me I should think to be my duty. But suppose that from the state of partial knowledge in which I think act *A* to be my duty, I could pass to a state of perfect knowledge in which I saw act *B* to be my duty, should I not say 'act *B* was the right act for me to do'? I should no doubt add 'though I am not to be blamed for doing act *A*'. But in adding this, am I not passing from the question 'what is right' to the question 'what is morally good'? At the same time I am not making the *full* passage from the one notion to the other; for in order that the act should be morally good, or an act I am not to be blamed for doing, it must not merely be the act which it is reasonable for me to think my duty; it must also be done for that reason, or from some other morally good motive. Thus the conception of the right act as the act which it is reasonable for me to think my duty is an unsatisfactory compromise between the true notion of the right act and the notion of the morally good action.

The general principles of duty are obviously not self-evident from the beginning of our lives. How do they come to be so? The answer is, that they come to be self-evident to us just as mathematical axioms do. We find by experience that this couple of matches and that couple make four matches, that this couple of balls on a wire and that couple make four balls; and by reflection on these and similar discoveries we come to see that it is of the nature of two and two to make four. In a precisely similar way, we see the *prima facie* rightness of an act which would be the fulfilment of a particular promise, and of another which would be the fulfilment of another promise, and when we have reached sufficient maturity to think in general terms, we apprehend *prima facie* rightness to belong to the nature of any fulfilment of promise. What comes first in time is the apprehension of the self-evident *prima facie* rightness of an individual act of a particular type. From this we come by reflection to apprehend the self-evident general principle of *prima facie* duty. From this, too, perhaps along with the apprehension of the self-evident *prima facie* rightness of the same act in virtue of its having another characteristic as well, and perhaps in spite of the apprehension of its *prima facie* wrongness in virtue of its having some third characteristic, we come to believe something not self-evident at all, but an object of probable opinion, viz. that this particular act is (not *prima facie* but) actually right.

In this respect there is an important difference between rightness and mathematical properties. A triangle which is isosceles necessarily has two of its angles equal, whatever other characteristics the triangle may have—whatever, for instance, be its area, or the size of its third angle. The equality of the two angles is a parti-resultant attribute.[12] And the same is true of all mathematical attributes. It is true, I may add, of *prima facie* rightness. But no act is ever, in virtue of falling under some general description, necessarily actually right; its rightness depends on its whole nature[13] and not on any element in it. The reason is that no mathematical object (no figure, for instance, or angle) ever has two characteristics that tend to give it opposite resultant characteristics, while moral acts often (as every one knows) and indeed always (as on reflection we must admit) have different characteristics that tend to make them at the same time *prima facie* right and *prima facie* wrong; there is probably no act, for instance, which does good to any one without doing harm to some one else, and vice versa.

·   ·   ·   ·   ·   ·   ·

It is worth while to try to state more definitely the nature of the acts that are right. We may try to state first what (if anything) is the universal nature of *all* acts that are right. It is obvious that any of the acts that we do has countless effects, directly or indirectly, on countless people, and the probability is that any act, however right it be, will have adverse effects (though these may be very trivial) on some innocent people. Similarly, any wrong act will probably have beneficial effects on some deserving people. Every act therefore, viewed in some aspects, will be *prima facie* right, and viewed in others, *prima facie* wrong, and right acts can be distinguished from wrong acts only as being those which, of all those possible for the agent in the circumstances, have the greatest balance of *prima facie* rightness, in those respects in which they are *prima facie* right, over their *prima facie* wrongness, in those respects in which they are *prima facie* wrong—*prima facie* rightness and wrongness being understood in the sense previously explained. For the estimation of the comparative stringency of these *prima facie* obligations no general rules can, so far as I can see, be laid down. We can only say that a great deal of stringency belongs to the duties of 'perfect obligation'—the duties of keeping our promises, of repairing wrongs we have done, and of returning the equivalent of services we have received. For the rest, ἐν τῇ αἰσθήσει ἡ κρίσις.[14] This sense of our particular duty in particular circumstances, preceded and informed by the fullest reflection we can bestow on the act in all its bearings, is highly fallible, but it is the only guide we have to our duty.

Notes

1. G.E. Moore, author of *Principia Ethica* (Cambridge: Cambridge University Press, 1903) and *Ethics* (London: Oxford University Press, 1911).—C.W.G.

2. I take the theory which, as I have tried to show, seems to be put forward in *Ethics* rather than the earlier and less plausible theory put forward in *Principia Ethica.* For the difference, cf. my pp. 8–11.

3. These are not strictly speaking duties, but things that tend to be our duty, or *prima facie* duties. Cf. pp. 19–20 [pp. 86–87 in this volume].

4. Hastings Rashdall, author of *The Theory of Good and Evil,* 2 vols., 2nd ed. (London: Oxford University Press, 1924).—C.W.G.

5. Some will think it, apart from other considerations, a sufficient refutation of this view to point out that I also stand in that relation to myself, so that for this view the distinction of oneself from others is morally insignificant.

6. H. A. Prichard, author of *Duty and Interest* (New York: Oxford University Press, 1928) and *Moral Obligation* (New York: Oxford University Press, 1950).—C.W.G.

7. I should make it plain at this stage that I am *assuming* the correctness of some of our main convictions as to *prima facie* duties, or, more strictly, am claiming that we *know* them to be true. To me it seems as self-evident as anything could be, that to make a promise, for instance, is to create a moral claim on us in someone else. Many readers will perhaps say that they do *not* know this to be true. If so, I certainly cannot prove it to them; I can only ask them to reflect again, in the hope that they will ultimately agree that they also know it to be true. The main moral convictions of the plain man seem to me to be, not opinions which it is for philosophy to prove or disprove, but knowledge from the start; and in my own case I seem to find little difficulty in distinguishing these essential convictions from other moral convictions which I also have, which are merely fallible

opinions based on an imperfect study of the working for good or evil of
certain institutions or types of action.

8. For a needed correction of this statement, cf. pp. 22–23 [pp. 88–89
in this volume].

9. Pp. 135–38.

10. But cf. the qualification in p. 33, note 2 [note 13 below in this
volume].

11. Cf. pp. 122–23.

12. Cf. pp. 28 [p. 94 in this volume] and 122–23.

13. To avoid complicating unduly the statement of the general view I
am putting forward, I have here rather overstated it. Any act is the origination of a great variety of things many of which make no difference to its
rightness or wrongness. But there are always many elements in its nature
(i.e. in what it is the origination of) that make a difference to its rightness
or wrongness, and no element in its nature can be dismissed without consideration as indifferent.

14. 'The decision rests with perception'. Arist. *Nicomachean Ethics,*
1109b 23, 1126b 4.

5

Moral Dilemmas

E. J. Lemmon

In this paper, I attempt to characterize different varieties of moral dilemma. An assumption made throughout is that an affirmative answer can be given to the question: does a human being have free will? Without this assumption, in fact, there does not seem to be much for ethics to be about.

There are very many different kinds of moral situation in which a human agent can find himself or put himself. Without making any pretense of defining the distinction between moral and nonmoral situations, let us merely list some kinds of situation which it would be generally agreed can safely be called moral. I shall begin with the most straightforward and gradually move into areas which could be described as "dilemmatic."

1

The first, and it seems the simplest, class of moral situation is this: we know what we are to do, or have to do, or ought to do, and sim-

From *The Philosophical Review* 70 (1962): 139–143 and 148–158. Reprinted with the permission of *The Philosophical Review*.

ply do it. Within this class, there are several subclasses, which it will be worth our while to distinguish, depending on the source of our knowledge of what we are to do. What sources are distinguished may well depend on the society to which the agent in question belongs. Thus, if I may stray, like so many philosophers, into the sociology of ethics for a while, our own society tends to distinguish such sources as duties, obligations, and moral principles (Classical Greek society, if we may go by its language, does not seem to have made a clear distinction between obligation and duty). For example, a soldier may receive a battle order, and act on it directly, because he knows that it is his *duty as a soldier* so to do. Or a man may know he is to attend a certain meeting, and do so, because, having given his word that he will be there, he is *under an obligation* to attend. Or a man may know that he ought to tell the truth, and do so, because he holds as a *moral principle* that one should always tell the truth—a slightly unrealistic example, since moral principles tend to be prohibitive rather than compelling: a better example would be that of a man who knows he is not to commit adultery with a certain woman, and does not do so, because he holds it to be a moral ruling that one should at no time commit adultery.

To summarize these three subcases: first, one may know what one is to do, and do it, because one knows it to be one's duty to do that thing; second, one may know what one is to do, and do it, because one knows oneself to be under an obligation to do that thing; third, one may know what one is to do, and do it, because one holds it to be the right thing to do in view of some moral rule.

It follows logically, I would wish to claim, that a man ought to do something, if it is his duty to do that thing. Equally, he ought to do it if he is under an obligation to do it, and he ought to do it if it is right, in view of some moral principle to which he subscribes, that he should do it. But the converse implications do not, I think, hold. It might be true that a man ought to do something, and yet it not be his duty to do it, because rather it is the case that he is under an obligation to do it; or, even though he ought to do it, he is under no obligation to do it, but rather it is his duty to do it; or, even though he ought to do it, it is not that it is right to do it in view of some moral principle which he holds, but rather a case of duty or obligation.

To see that these converse implications fail, it will be necessary to take a closer look at our (rather parochial) concepts of duty and obligation. A man's duties are closely related to his special status or position. It nearly always makes sense to ask of a duty "duty *as what?*" The most straightforward case is that of duties incurred in virtue of a job: thus one has duties as a policeman, duties as headmaster, duties as prime minister or garbage-collector. In many societies, family relationships are recognized as determining duties: thus there are duties as a father, mother, son, or daughter. Less clearly delineated duties, in our society at least, are those of a host, those of a friend, those of a citizen. I do not think there are such things as one's duties as a human being, unless they be duties toward dogs or other members of the animal kingdom, for being a human being is not being in any special or distinguishing position, unless it is vis-à-vis dogs perhaps. The same point emerges from the adjective "dutiful." A dutiful *X* is someone who does his duties *as an X;* a dutiful parent is one who does his duty as a parent. No clear sense attaches to the phrase "a dutiful bachelor," at least in our society, for the status of bachelor is not thought of as bringing with it certain duties.

If duties are related to a special position or status, which distinguishes the man holding the position or status from others, obligations on the other hand are typically incurred by previous committing actions. Of course, again what actions are regarded as committal will vary from society to society. To us, the most familiar committing actions are promising or giving one's word generally, and signing one's signature. If you swear to tell the truth, from the moment of swearing you are under an obligation to tell the truth. If you promise to attend a meeting, then from that moment you are under an obligation to attend a meeting. If you sign your name to an IOU, then from that moment you are under an obligation to return the borrowed money. Less clearly delineated cases of obligations, at least in our society, are the obligation to return hospitality having received it and the obligation to give money to a beggar having been asked for it. This last case illustrates a concept which has relatively rare application for us—that of being put under an obligation to someone by their conduct rather than one's own. In certain societies, I believe, a knock on the door of one's house by a stranger at once puts one under an obligation of a firm kind to provide hospitality and, if necessary, a bed for the night.

If this admittedly sketchy analysis of the notions of duty and obligation is at all correct, it becomes easy to see how a man can be under an obligation to do something, though it is not his duty to do it, or how a man's duty may be to do something though he is under no obligation to do it. For example, it may be true that I ought to vote against a Communist candidate in some election, because it is my duty as a citizen to do this, though there is no clear sense in which I am under an obligation to vote against the Communist (I have made no promises, accepted no bribes, given my word in advance to no one). On the other hand, it could easily be that I am, in a different situation, under an obligation to vote against the Communist just because I have given my word that I shall do so, even though it may not in fact be my duty to do so. This is not, of course, to deny that we may both be under an obligation and have it as a duty to do something. For example, in the witness stand it is my duty as a witness to tell the truth, and I am also under an obligation to tell the truth since I have sworn an oath to do so.

An interesting borderline case between obligation and duty which when properly understood helps, I think, to mark the watershed between them, is the following. Children are often thought to be under some kind of obligation to help their parents in old age, and it is often thought that it is their duty to do so. Is this more properly to be considered a case of obligation or a case of duty? I suggest that we can consider it both ways, but that thinking of it one way is different from thinking of it the other way. If we regard it as a duty to help our parents, we are thinking rather of our special relationship to them, our status as children. If, on the other hand, we think of ourselves as under an obligation to our parents, it is surely in virtue of what they have done for us in the past, when we were children, that we are under this obligation—that is, it will be a case of our having been put under an obligation in some way by them. This difference in the mode of thought becomes clear if we vary the example slightly. Suppose they turn out to be not parents but foster parents. Then we may well feel that our duty is less because the relationship is less close, but our sense of obligation may be no less great in view of what they have done for us. On the other hand, if our parents have not in point of fact done a great deal for us, we may feel in no sense under any obligation to help them, but our sense of duty may be just as real because of our close relationship with them.

Broadly speaking, then, duty-situations are status-situations while obligation-situations are contractual-situations. Both duties and obligations may be sources of "ought's," but they are logically independent sources. And a third source, independent of the other two, is that it is right to do something in view of a moral principle.

I have not discussed this here because it is well discussed in almost all contemporary ethical writing, while the concepts of duty and obligation tend to be neglected.

I shall not in fact be very disturbed to learn that there are aspects of the concept of duty or the concept of obligation which I have omitted, or even that I have missed either concept's most central aspect, as we have it. My main concern is rather that there are generically different ways in which it can come to be true that we ought to do something or ought not to do something. While "ought" is a very general word of ethical involvement, "duty" and "obligation" and "right," as I am using them at least, are highly specialized words.

Yet another way of putting what I want to say would be the following. It is analytic that one ought to do one's duty and that one ought to fulfill one's obligations. But that it is one's duty to do what one ought to do and that one is under an obligation to do what one ought to do are synthetic and false. In the case of "right," I think it is analytic that one ought to do what it is right to do, but I am not sure whether it is analytic or synthetic that it is right to do what one ought to do—for special reasons which I will not go into here.

.   .   .   .

3

It is well past time to reach the main topic of this paper. My third class of moral situation constitutes what I take to be the simplest variety of moral dilemma in the full sense. The characterization of this class is as follows: a man both ought to do something and ought not to do that thing. Here is a simple example, borrowed from Plato. A friend leaves me with his gun, saying that he will be back for it in the evening, and I promise to return it when he calls. He arrives in a distraught condition, demands his gun, and announces that he is going to shoot his wife because she has been unfaithful. I ought to return the gun, since I promised to do so—a case of obligation. And yet I ought not to do so, since to do so would be to be indirectly responsible for a murder, and my moral principles are such that I regard this as wrong. I am in an extremely straightforward moral dilemma, evidently resolved by not returning the gun.

The description of this class of cases may perhaps cause alarm;

for it may well be thought to be contradictory that a man both ought and ought not to do something. To indicate why I do not think this is so, I will begin by considering the logic of the modal verb "has to," or "must," and then contrast this logic with that of the modal "ought." If a man has to do something,[1] it does follow that he does that thing, in the sense that if he does not do it it cannot have been true that he must do it. This emerges quite clearly, I think, from the following fact of usage: a man announces that he must do something, but it later emerges that he has not done that thing. Then he will now repeat his earlier claim, not in the form that he *had* to do it, which would suggest falsely that he had done it, but in the form that he *ought* to have done it. Or, if at a party I say that I have to go, this will be taken as a sign heralding my departure. But if I merely say that I ought to go, this is entirely compatible, human weakness being what it is, with my staying for another hour.

It follows from this that "must" and "must not" are contraries in the logician's sense. That is, it cannot be both true that I must do something and that I must not do it: for if I must I will, and if I must not I will not, which is a contradiction. On the other hand, it may well be the case that I neither must do something nor must not do that thing. For example, it is neither true that I must light a cigarette nor that I must not light a cigarette. Hence "must" and "must not" may well both be false, though they may not be both true.

"Must not" should not be supposed to be the negation of "must," in the way that "cannot" is the negation of "can." The proper negation of "I must tell the truth" is, roughly, "I do not have to tell the truth." In an entirely similar way, "ought not" is certainly not the negation of "ought." For example, it is true neither that I ought to be playing chess nor that I ought not to be playing chess; hence "ought" and "ought not" are not contradictory. But are they even contrary to one another? In the case of "must" and "must not" we showed them to be contrary by showing that "must" implied "will" and that consequently "must not" implied "will not." Hence an explicit contradiction is derivable from the assumption that a man both must and must not do something. But no similar contradiction is derivable from the assumption that someone both ought and ought not to do something; for it certainly does not follow from the fact that a man ought to do something that he will do it, nor does it follow from the fact that a man ought not to do something that he will not do that thing. There seems no reason, therefore, why we should regard "ought" and "ought not" even as contraries, still less as contradictories. It seems to me that "ought" and "ought not" may well both be true, and that this description in fact characterizes a certain class of moral dilemma. Indeed, the Platonic example cited would not be a dilemma at all unless it was true that the man both ought to return the gun and ought not to return it. It is a nasty fact about human life that we sometimes both ought and ought not to do things; but it is not a logical contradiction.[2]

My motive for carefully distinguishing some of the sources for "ought's" earlier in this paper should now be apparent. For moral dilemmas of the sort we are at present considering will appear generally[3] in the cases where these sources conflict. Our duty may conflict with our obligations, our duty may conflict with our moral principles, or our obligations may conflict with our moral principles. The Platonic case was an example of a conflict between principle and obligation. A simple variant illustrates a conflict between obligation and duty: the man with whom the gun is deposited may regard it as his duty as a friend not to return the gun, even though he is under an obligation to do so. And duty conflicts with principle every time that we are called on in our jobs to do things which we find morally repugnant.

A natural question to ask next is: how are moral dilemmas of this simple kind to be resolved? There are certain very simple resolutions, known from the philosophical literature, which we should discuss first; but I do not think they are in practice very common. First, we may hold to some very sweeping "higher-order principle" such as "Always prefer duty to obligation" or "Always follow moral principles before duty or obligation." This last precept, for example, at once resolves the Platonic dilemma mentioned earlier, which, as I described it, was a simple clash between principle and obligation. Secondly, and rather less simply, we may have in advance a complex ordering of our various duties, obligations, and the like—putting, for example, our duties as a citizen before our duties as a friend and our duties as a friend before any obligations we may have incurred—in virtue of which the moral dilemma is resolved. But dilemmas in which we are morally prepared, in which we, as it were, merely have to look up the solution in our private ethical code, are rare, I think, and in any case of little practical interest. Of greater importance are those dilemmas in this class where some decision of a moral character is required. And here it must be remembered that the failure to make a decision in one sense is itself to make a decision in another, broader, sense. For our predicament is here so described that, whatever we do, even if we do nothing at all (whatever that might mean), we are doing something which we ought not to do, and so can be called upon to justify either our activity or our inactivity. The only way we can avoid a decision is by ceasing to be any longer an agent (e.g., if we are arrested, or taken prisoner, or kidnapped, or die). This precise situation leads to another familiar pattern of bad faith, in which we pretend to ourselves either that no decision is called for or that in one way or another the decision has been taken out of our hands by others or that we are simply the victims of our own character in acting in this way or that, that we cannot help doing what we do do and so cannot be reproached for resolving the dilemma in this way or that. If, however, we are to act here in good faith, we shall recognize that the dilemma is what it is and make the best decision we can.

Now what kind of considerations may or should affect the decision? The situation is such that no moral, or at least purely moral, considerations are relevant, in the sense that no appeal to our own given morality can decide the issue. We may of course consult a friend, take moral advice, find out what others have done in similar situations, appeal as it were to precedent. But again none of these appeals will be decisive—we still have to decide to act in accordance with advice or precedent. Or again we may approach our decisions by a consideration of ends—which course of action will, so far as we can see, lead to the best result. (I do not think it is an accident, by the way, that the word "good," or rather its superlative "best," makes its first appearance at this point in our discussion; for it is typically when we are torn between courses of conduct that the question of comparing different actions arises, and hence the word "good," a comparative adjective unlike "right," is at home here; the consequence, admittedly paradoxical, of this view of "good" is that it is not properly a word of moral appraisal at all, despite the vast attention it receives from ethical philosophers; and I think I accept this conclusion). Thus a consideration of ends determines a solution to the Platonic dilemma discussed earlier. Although I ought to return the gun and also ought not to return the gun, in fact it is evidently best, when we weigh up the expected outcome, not to return the gun, and so to sacrifice one's obligation to utilitarian considerations. Of course, when I say that this solution is evidently the best, I do not mean that it cannot be questioned. What I do mean is that it can only be seriously questioned by someone whose whole attitude toward human life is basically different from that of a civilized Western human being. Someone who thinks that it would really be better to return the gun must either hold the importance of a man's giving his word to be fantastically high or else hold human life to be extremely cheap, and I regard both these attitudes as morally primitive.

4

I shall pass on now to the next, more complex, class of moral situations which might be described as dilemmatic in the full sense. Roughly, the class I now have in mind may be described thus: there is some, but not conclusive, evidence that one ought to do something, and there is some, but not conclusive, evidence, that one ought not to do that thing.[4] All the difficulties that arose in the way of making a decision in the last class of cases arise typically here too, but there are now difficulties of a new kind as well. Moreover, in this class of cases there can be no preassigned moral solution to the dilemma in virtue of higher-order principles or a given ordering of one's duties and obligations and the like, because part of the very dilemma is just one's uncertainty as to one's actual moral situation, one's situation with respect to duties, obligations, and principles. For example, it may be unclear whether it really is one's duty as a citizen to vote against the Communist candidate, and also unclear whether one is under an obligation to vote for the Communist candidate in view, let us say, of financial help received from the Communists in the Resistance during the war. Hence one is in a moral dilemma because there is some evidence that one should vote Communist and some that one should not, though in neither case is the evidence conclusive.

A good illustration of the kind of complexity this type of situation may embrace is . . . from Sartre:

> I will refer to the case of a pupil of mine who sought me out in the following circumstances. His father was quarrelling with his mother and was also inclined to be a "collaborator"; his elder brother had been killed in the German offensive of 1940 and this young man, with a sentiment somewhat primitive but generous, burned to avenge him. His mother was living alone with him, deeply afflicted by the semitreason of his father and by the death of her oldest son, and her one consolation was in this young man. But he, at this moment, had the choice between going to England to join the Free French Forces

or of staying near his mother and helping her to live. He fully realized that this woman lived only for him and that his disappearance—or perhaps his death—would plunge her into despair. He also realized that, concretely and in fact, every action he performed on his mother's behalf would be sure of effect in the sense of aiding her to live, whereas anything he did in order to go and fight would be an ambiguous action which might vanish like water into sand and serve no purpose. For instance, to set out for England he would have to wait indefinitely in a Spanish camp on the way through Spain; or, on arriving in England or in Algiers he might be put into an office to fill up forms. Consequently, he found himself confronted by two very different modes of action; the one concrete, immediate, but directed towards only one individual; the other an action addressed to an end infinitely greater, a national collectivity, but for that reason ambiguous—and it might be frustrated on the way. At the same time, he was hesitating between two kinds of morality; on the one side, the morality of sympathy, of personal devotion and, on the other side, a morality of wider scope but of more debatable validity. He had to choose between these two.[5]

A crude oversimplification of this example might depict it thus: the boy is under some obligation to stay with his mother; or, perhaps better, his mother by her own position has put him under some obligation to stay with her, since she is now dependent on him for her own happiness. Consequently, he is conscious in some degree that he ought to stay with her. On the other hand he feels some kind of duty to join the Free French in England—a duty perhaps to his country as a citizen. But this duty is far from being clearly given; as Sartre stresses, it is felt only ambiguously. It may be his duty to fight, but can it really be his duty, given his obligation to his mother, to sit in an office filling out forms? He is morally torn, but each limb of the moral dilemma is not itself here clearly delineated.

An interesting feature of this case, and of the class of cases in general which we are considering, is that, in attempting to reach a decision, the arguments which try to establish exactly what one's moral situation is are not distinguishable from those which attempt to resolve the dilemma itself. Thus the boy is unclear where his duty lies partly because he is unclear what exactly would be the outcome of his decision to leave his mother, and this outcome is also relevant to the decision itself, as a utilitarian consideration affecting his choice.

Sartre's example has an important further feature, which marks out a particular subclass of the class of moral dilemmas in general:

the dilemma is so grave a one, personally speaking, that either decision in effect marks the adoption on the part of the agent of a changed moral outlook. It does not seem to have been much observed by ethical philosophers that, speaking psychologically, the adoption of a new morality by an agent is frequently associated with the confrontation of a moral dilemma. Indeed, it is hard to see what else would be likely to bring about a change of moral outlook other than the having to make a difficult moral decision. On the nature of such a change there is time here only to say a few things. First, the change frequently and always in serious cases is associated with a change in fundamental attitudes, such as the change from liberalism to conservatism in politics or the change from Christianity to atheism in the field of religion. And the reasons given for the moral change may well be identical with the reasons given for the change in fundamental attitudes. This last kind of change is neither fully rational nor fully irrational. To persuade someone to change his fundamental attitudes is like getting someone to see an aesthetic point—to appreciate classical music or impressionist painting, for example. Arguments can be given, features of music or painting may be drawn to the person's attention, and so on and so forth, but none of these reasons is finally conclusive. Nonetheless, we should not rush to the opposite conclusion that matters of aesthetic taste are purely subjective. In a somewhat similar way we may persuade someone, or he may persuade himself, to change his fundamental attitudes, and so to change his moral outlook, at a time of moral crisis. Roughly speaking, Sartre's boy has to decide whether to be politically engaged or not, and this decision may well affect and be affected by his fundamental attitudes.

I am not at all saying that this kind of serious case is common; indeed, I think it is rare; but it is still of the greatest importance to ethics to investigate it, because it is of the greatest practical importance in a man's life. There may well be people who have never had to face a moral situation of these dimensions. But for Antigones and others who live faced with occasional major crises, the appropriate reasoning for this kind of moral dilemma is of vital importance. On the other hand, it is not at all clear what the role of the philosopher should be here. If we listen to much of contemporary ethical writing, his role is merely to analyze the discourse in which such reasoning is couched; the task of deciding what are good and what are bad ethical arguments belongs to someone else, though it is never quite made clear to whom. It is my own view that, even though it may be part and an important part of the philosopher's job to analyze the terminology of ethical arguments, his job does not stop there. Perhaps no one is properly equipped to give moral advice to anyone else, but if anyone is it is the philosopher, who at least may be supposed to be able to detect bad reasoning from good. It is a corollary of this view that a philosopher is not entitled to a private life—by which I mean that it is his duty to hold political and religious convictions in such a form as to be philosophically defensible or not to hold them at all. He is not entitled to hold such beliefs in the way in which many nonphilosophers hold them, as mere articles of faith.

5

After this brief digression, I must return to my classification of moral dilemmas: for there is one more kind that, with some hesitation, I should like to introduce. This is an even more extreme kind of dilemma than the last and probably of even rarer occurrence. I mean the kind of situation in which an agent has to make a decision of a recognizably moral character though he is completely unprepared for the situation by his present moral outlook. This case differs from the last in that there the question was rather of the applicability of his moral outlook to his present situation, while here the question is rather how to create a new moral outlook to meet unprecedented moral needs. This case is in some respects easier for the agent and in some respects harder to face than the last: easier, if he recognizes the situation for what it is, because he at least knows that for sure he has some basic moral rethinking to do, which is often not clear in the previous case; but harder, because basic moral rethinking is harder work in general than settling the applicability of given moral principles to a particular situation. A typical, but morally wrong, way of escape from this dilemma is again to act in bad faith, by pretending to oneself that the situation is one which one can handle with one's given moral apparatus.

A possible real instance of this kind of moral dilemma is that which faced Chamberlain in his negotiations with Hitler in 1938. He ought to have realized that he was dealing with a kind of person for which his own moral outlook had not prepared him, and that as prime minister he was called upon to rethink his moral and political approach in a more realistic way. This he failed to do, either because he was genuinely deceived as to Hitler's real character or, as I suspect, because he deceived himself on this point: if the latter, then he was guilty of the type of bad faith to which I am alluding.

The main point of this variety of moral dilemma is that, at least if correctly resolved, it forces a man to develop a new morality; in the case of the last type of dilemma, this was a possible outcome but by no means a necessary one. So perhaps this is the place at which to say a little about what is involved in such a development. Here the analogy with aesthetics, which Sartre and others have cautiously drawn, may be useful. There may come a point in the development of a painter, say, or a composer, where he is no longer able to go on producing work that conforms to the canons of composition which he has hitherto accepted, where he is compelled by his authenticity as a creator to develop new procedures and new forms. It is difficult to describe what will guide him in the selection of new canons, but one consideration will often be the desire to be (whatever this means) *true to himself*. It may well be that an appropriate consideration in the development of a new moral outlook is the desire to be, in the relevant sense whatever that is, true to oneself and to one's own character. But I will not pursue this topic here, because I confess myself to be quite in the dark as to what the sense of these words is.

To conclude, I will not attempt to summarize, but rather I will say what I would like to see done and what I know I have here failed properly to do. I should like to see a detailed breakdown of the different kinds of difficult moral situations in which human beings, living as they do in societies, find themselves, because in my opinion too much attention has been paid in contemporary ethical writing to the easy, rule-guided, moral situation. The five types of ethical situation which I have here tried to distinguish might well be replaced by five hundred types, human life being what it is. Such an analysis will require sympathetic treatment of real moral problems considered in detail, and it will require a proper analysis of the concepts of choice and decision—active moral concepts, rather than the passive, spectatorlike, concepts of good and right. Secondly, I should like to see a proper discussion of the arguments that go to resolve moral dilemmas, because I do not believe that this is an area of total irrationality, though I do not believe that a traditional logical approach (the logic of imperatives, deontic logic, and whatnot) will do either. This will entail saying what constitutes a good and a bad moral reason for making a moral decision, and so will bring the moral philosopher out from his corner, where I think he has been

too long, and back into the familiar but forgotten Socratic position
of trying to answer the ever-present but ever-changing question:
how should a man live?

Notes

1. As opposed, I think, to *having it to do:* for that might mean only that
it was on his agenda, i.e., list of things he ought to do.

2. Professor M. Lazerowitz pointed out to me in discussion the following consequence of the view above: if X ought to do P and ought to do Q,
then X ought to do P and Q, by a principle of deontic logic which I and
others accept; hence, in the cases under consideration, X ought to do both
P and not-P, and yet it is a logical truth that X cannot do both P and not-
P; so a contradiction seems to be obtained. I view this, however, as a refutation of the principle that "ought" implies "can," to which there are surely
clear counterexamples even without the introduction of the present
instances.

3. Though of course by no means always: two duties, or two obligations, may well conflict with one another.

4. At this point we bid farewell to the deontological mapping of moral
concepts on which we have partially relied up to now. The new area is not
charted enough for that and perhaps should not be charted in that way at
all.

5. Sartre, *Existentialism and Humanism,* trans. by P. Mairet (London,
1948), pp. 35–36.

6

Ethical Consistency

Bernard Williams

I shall not attempt any discussion of ethical consistency in general. I shall consider one question that is near the centre of that topic: the nature of moral conflict. I shall bring out some characteristics of moral conflict that have bearing, as I think, on logical or philosophical questions about the structure of moral thought and language. I shall centre my remarks about moral conflict on certain comparisons between this sort of conflict, conflicts of beliefs, and conflicts of desires; I shall start, in fact, by considering the latter two sorts of conflict, that of beliefs very briefly, that of desires at rather greater length, since it is both more pertinent and more complicated.

Some of what I have to say may seem too psychological. In one respect, I make no apology for this; in another, I do. I do not, in as much as I think that a neglect of moral psychology and in particular of the role of emotion in morality has distorted and made unrealistic a good deal of recent discussion; having disposed of emotivism

From *Problems of the Self: Philosophical Papers 1956–1972* (Cambridge: Cambridge University Press, 1973), pp. 166–86. Copyright © Cambridge University Press. Reprinted with the permission of the publisher and the author. An earlier version of this essay appeared in *Proceedings of the Aristotelian Society* supp. vol. 39 (1965): 103–24.

as a theory of the moral judgement, philosophers have perhaps tended to put the emotions on one side as at most contingent, and therefore philosophically uninteresting, concomitants to other things which are regarded as alone essential. This must surely be wrong: to me, at least, the question of what emotions a man feels in various circumstances seems to have a good deal to do, for instance, with whether he is an admirable human being or not. I do apologise, however, for employing in the following discussion considerations about emotion (in particular, *regret*) in a way which is certainly less clear than I should like.

1

It is possible for a man to hold inconsistent beliefs, in the strong sense that the statements which would adequately express his beliefs involve a logical contradiction. This possibility, however, I shall not be concerned with, my interest being rather in the different case of a man who holds two beliefs which are not inconsistent in this sense, but which for some empirical reason cannot both be true. Such beliefs I shall call 'conflicting'. Thus a man might believe that a certain person was a minister who took office in October 1964 and also that that person was a member of the Conservative party. This case will be different from that of inconsistent beliefs, of course, only if the man is ignorant of the further information which reveals the two beliefs as conflicting, viz. that no such minister is a Conservative. If he is then given this information, and believes it, then either he becomes conscious of the conflict between his original beliefs[1] or, if he retains all three beliefs (for instance, because he has not 'put them together'), then he is in the situation of having actually inconsistent beliefs. This shows a necessary condition of beliefs conflicting: that if a pair of beliefs conflict, then (a) they are consistent and (b) there is a true factual belief which, if added to the original pair, will produce a set that is inconsistent.

2

What is normally called conflict of *desires* has, in many central cases, a feature analogous to what I have been calling conflict of beliefs: that the clash between the desires arises from some contingent matter of fact. This is a matter of fact that makes it impossible for both the desires to be satisfied; but we can consistently imagine a state of affairs in which they could both be satisfied. The contingent root of the conflict may, indeed, be disguised by a use of language that suggests logical impossibility of the desires being jointly satisfied; thus a man who was thirsty and lazy, who was seated comfortably, and whose drinks were elsewhere, might perhaps represent his difficulty to himself as his both wanting to remain seated and wanting to get up. But to put it this way is for him to hide the roots of his difficulty under the difficulty itself; the second element in the conflict has been so described as to reveal the obstacle to the first, and not its own real object. The sudden appearance of help, or the discovery of drinks within arm's reach, would make all plain.

While many cases of conflict of desires are of this contingent character, it would be artificial or worse to try to force all cases into this mould, and to demand for every situation of conflict an answer to the question 'what conceivable change in the contingent facts of the world would make it possible for both desires to be satisfied?' Some cases involving difficulties with space and time, for instance, are likely to prove recalcitrant: can one isolate the relevant contingency in the situation of an Australian torn between spending Christmas in Christmassy surroundings in Austria, and spending it back home in the familiar Christmas heat of his birthplace?

A more fundamental difficulty arises with conflicts of desire and aversion towards one and the same object. Such conflicts can be represented as conflicts of two desires: in the most general case, the desire to have and the desire not to have the object, where 'have' is a variable expression which gets a determinate content from the context and from the nature of the object in question.[2] There are indeed other cases in which an aversion to x does not merely take the form of a desire *not to have* x (to avoid it, reject it, to be elsewhere, etc.), but rather the form of a desire that x *should not exist*— in particular, a desire to destroy it. These latter cases are certainly different from the former (aversion here involves advancing rather than retreating), but I shall leave these, and concentrate on the former type. Conflicts of desire and aversion in this sense differ from the conflicts mentioned earlier, in that the most direct characterization of the desires—'I want to have x' and 'I want not to have x'— do not admit an imaginable contingent change which would allow both the desires to be satisfied, the descriptions of the situations that would satisfy the two desires being logically incompatible. However, there is in many cases something else that can be imagined which is just as good: the removal from the object of the disadvantageous features which are the ground of the aversion or (as I shall call aversions which are merely desires *not to have*) negative desire. This imaginable change would eliminate the conflict, not indeed by satisfying, but by eliminating, the negative desire.

This might be thought to be cheating, since any conflict of desires can be imagined away by imagining away one of the desires. There is a distinction, however, in that the situation imagined without the negative desire involves no loss of utility: no greater utility can be attached to a situation in which a purely negative desire is satisfied, than to one in which the grounds of it were never present at all. This does not apply to desires in general (and probably not to the more active, destructive, type of aversion distinguished before). Admittedly, there has been a vexed problem in this region from antiquity on, but (to take the extreme case) it does seem implausible to claim that there is no difference of utility to be found between the lives of two men, one of whom has no desires at all, the other many desires, all of which are satisfied.

Thus it seems that for many cases of conflict of desire and aversion towards one object, the basis of the conflict is still, though in a slightly different way, contingent, the contingency consisting in the coexistence of the desirable and the undesirable features of the object. Not all cases, however, will yield to this treatment, since there may be various difficulties in representing the desirable and undesirable features as only contingently coexisting. The limiting case in this direction is that in which the two sets of features are identical (the case of ambivalence)—though this will almost certainly involve the other, destructive, form of aversion.

This schematic discussion of conflicts between desires is meant to apply only to nonmoral desires; that is to say, to cases where the answer to the question 'why do you want x?' does not involve expressing any moral attitude. If this limitation is removed, and moral desires are considered, a much larger class of noncontingently based conflicts comes into view, since it is evidently the case that a moral desire and a nonmoral desire which are in conflict may be directed towards exactly the same features of the situation.[3] Leaving moral desires out of it, however, I think we find that a very large range of conflicts of desires have what I have called a contingent basis. Our desires that conflict are standardly like beliefs that conflict, not like beliefs that are inconsistent; as with conflicting beliefs it is the world, not logic, that makes it impossible for them both to be true, so with most conflicting desires, it is the world, not logic, that makes it impossible for them both to be satisfied.

3

There are a number of interesting contrasts between situations of conflict with beliefs and with desires; I shall consider two.

(*a*) If I discover that two of my beliefs conflict, at least one of them, by that very fact, will tend to be weakened; but the discovery that two desires conflict has no tendency, in itself, to weaken either of them. This is for the following reason: while satisfaction is related to desire to some extent as truth is related to belief, the discovery that two desires cannot both be satisfied is not related to those desires as the discovery that two beliefs cannot both be true is related to those beliefs. To believe that *p* is to believe that *p* is true, so the discovery that two of my beliefs cannot both be true is itself a step on the way to my not holding at least one of them; whereas the desire that I should have such-and-such, and the belief that I will have it, are obviously not so related.

(*b*) Suppose the conflict ends in a decision, and, in the case of desire, action; in the simplest case, I decide that one of the conflicting beliefs is true and not the other, or I satisfy one of the desires and not the other. The rejected belief cannot substantially survive this point, because to decide that a belief is untrue *is* to abandon, i.e., no longer to have, that belief. (Of course, there are qualifications to be made here: it is possible to say 'I know that it is untrue, but I can't help still believing it.' But it is essential to the concept of belief that such cases are secondary, even peculiar.) A rejected desire, however, can, if not survive the point of decision, at least reappear on the other side of it on one or another guise. It may reappear, for instance, as a general desire for something of the same sort as the object rejected in the decision; or as a desire for another particular object of the same sort; or—and this is the case that will concern us most—if there are no substitutes, the opportunity for satisfying that desire having irrevocably gone, it may reappear in the form of a *regret* for what was missed.

It may be said that the rejection of a belief may also involve regret. This is indeed true, and in more than one way: if I have to abandon a belief, I may regret this either because it was a belief of

mine (as when a scientist or a historian loses a pet theory), or—
quite differently—because it would have been more agreeable if the
world had been as, when I had the belief, I thought it was (as when
a father is finally forced to abandon the belief that his son survived
the sinking of the ship). Thus there are various regrets possible for
the loss of beliefs. But this is not enough to reinstate a parallelism
between beliefs and desires in this respect. For the regret that can
attach to an abandoned belief is never sufficiently explained just by
the fact that the man did have the belief; to explain this sort of
regret, one has to introduce something else—and this is, precisely,
a desire, a desire for the belief to be true. That a man regrets the
falsification of his belief that p shows not just that he believed that
p, but that he wanted to believe that p: where 'wanting to believe
that p' can have different sorts of application, corresponding to the
sorts of regret already distinguished. That a man regrets not having
been able to satisfy a desire is sufficiently explained by the fact that
he had that desire.

4

I now turn to moral conflict. I shall discuss this in terms of *ought,*
not because *ought* necessarily figures in the expression of every
moral conflict, which is certainly not true, but because it presents
the most puzzling problems. By 'moral conflict' I mean only cases
in which there is a conflict between two moral judgements that a
man is disposed to make relevant to deciding what to do; that is to
say, I shall be considering what has traditionally, though mislead-
ingly, been called 'conflict of obligations', and not, for instance, conflicts between a moral judgement and a nonmoral desire, though
these, too, could naturally enough be called 'moral conflicts'. I shall
further omit any discussion of the possibility (if it exists) that a man
should hold moral principles or general moral views which are
intrinsically inconsistent with one another, in the sense that there
could be no conceivable world in which anyone could act in accordance with both of them; as might be the case, for instance, with a
man who thought both that he ought not to go in for any blood-
sport (as such) and that he ought to go in for foxhunting (as such).
I doubt whether there are any interesting questions that are peculiar
to this possibility. I shall confine myself, then, to cases in which the
moral conflict has a contingent basis, to use a phrase that has
already occurred in the discussion of conflicts of desires.

Some real analogy, moreover, with those situations emerges if one considers two basic forms that the moral conflict can take. One is that in which it seems that I ought to do each of two things, but I cannot do both. The other is that in which something which (it seems) I ought to do in respect of certain of its features also has other features in respect of which (it seems) I ought not to do it. This latter bears an analogy to the case of desire and aversion directed towards the same object. These descriptions are of course abstract and rather artificial; it may be awkward to express in many cases the grounds of the *ought* or *ought not* in terms of features of the thing I ought or ought not to do, as suggested in the general description. I only hope that the simplification achieved by this compensates for the distortions.

The two situations, then, come to this: in the first, it seems that I ought to do *a* and that I ought to do *b,* but I cannot do both *a* and *b;* in the second, it seems that I ought to do *c* and that I ought not to do *c.* To many ethical theorists it has seemed that actually to accept these seeming conclusions would involve some sort of logical inconsistency. For Ross, it was of course such situations that called for the concept of *prima facie* obligations: two of these are present in each of these situations, of which at most one in each case can constitute an actual obligation. On Hare's views,[4] such situations call (in some logical sense) for a revision or qualification of at least one of the moral principles that give rise, in their application, to the conflicting *ought's.* It is the view, common to these and to other theorists, that there is a logical inconsistency of some sort involved here, that is the ultimate topic of this paper.

5

I want to postpone, however, the more formal sorts of consideration for a while, and try to bring out one or two features of what these situations are, or can be, like. The way I shall do this is to extend further the comparison I sketched earlier, between conflicts of beliefs and conflicts of desires. If we think of it in these terms, I think it emerges that there are certain important respects in which these moral conflicts are more like conflicts of desires than they are like conflicts of beliefs.

(*a*) The discovery that my factual beliefs conflict *eo ipso* tends to weaken one or more of the beliefs; not so, with desires; not so, I think, with one's conflicting convictions about what one ought to do. This comes out in the fact that conflicts of *ought's,* like conflicts of desires, can readily have the character of a struggle, whereas conflicts of beliefs scarcely can, unless the man not only believes these things, but wants to believe them. It is of course true that there are situations in which, either because of some practical concern connected with the beliefs, or from an intellectual curiosity, one may get deeply involved with a conflict of beliefs, and something rather like a struggle may result: possibly including the feature, not uncommon in the moral cases, that the more one concentrates on the dilemma, the more pressing the claims of each side become. But there is still a difference, which can be put like this: in the belief case my concern to get things straight is a concern both to find the right belief (whichever it may be) and to be disembarrassed of the false belief (whichever it may be), whereas in the moral case my concern is not in the same way to find the right item and be rid of the other. I may wish that the facts had been otherwise, or that I had never got into the situation; I may even, in a certain frame of mind, wish that I did not have the moral views I have. But granted that it is all as it is, I do not think in terms of banishing error. I think, if constructively at all, in terms of acting for the best, and this is a frame of mind that *acknowledges* the presence of both the two *ought's.*

(*b*) If I eventually choose for one side of the conflict rather than the other, this is a possible ground of regret—as with desires, although the regret, naturally, is a different sort of regret. As with desires, if the occasion is irreparably past, there may be room for nothing but regret. But it is also possible (again like desires) that the moral impulse that had to be abandoned in the choice may find a new object, and I may try, for instance, to 'make up' to people involved for the claim that was neglected. These states of mind do not depend, it seems to me, on whether I am convinced that in the choice I made I acted for the best; I can be convinced of this, yet have these regrets, ineffectual or possibly effective, for what I did not do.

It may be said that if I am convinced that I acted for the best; if, further, the question is not the different one of self-reproach for having got into the conflict-situation in the first place; then it is merely irrational to have any regrets. The weight of this comment depends on what it is supposed to imply. Taken most naturally, it implies that these reactions are a bad thing, which a fully admirable moral agent (taken, presumably, to be rational) would not display. In this sense, the comment seems to me to be just false. Such reactions do not appear to me to be necessarily a bad thing, nor an agent who displays them *pro tanto* less admirable than one who does not. But I do not have to rest much on my thinking that this is so; only on the claim that it is not inconsistent with the nature of morality to think that this is so. This modest claim seems to me undeniable. The notion of an admirable moral agent cannot be all that remote from that of a decent human being, and decent human beings are disposed in some situations of conflict to have the sort of reactions I am talking about.

Some light, though necessarily a very angled one, is shed on this point by the most extreme cases of moral conflict, tragic cases. One peculiarity of these is that the notion of 'acting for the best' may very well lose its content. Agamemnon at Aulis may have said 'May it be well,'[5] but he is neither convinced nor convincing. The agonies that a man will experience after acting in full consciousness of such a situation are not to be traced to a persistent doubt that he may not have chosen the better thing; but, for instance, to a clear conviction that he has not done the better thing because there was no better thing to be done. It may, on the other hand, even be the case that by some not utterly irrational criteria of 'the better thing', he is convinced that he did the better thing: rational men no doubt pointed out to Agamemnon his responsibilities as a commander, the many people involved, the considerations of honour, and so forth. If he accepted all this, and acted accordingly: it would seem a glib moralist who said, as some sort of criticism, that he must be irrational to lie awake at night, having killed his daughter. And he lies awake, not because of a doubt, but because of a certainty. Some may say that the mythology of Agamemnon and his choice is nothing to us, because we do not move in a world in which irrational gods order men to kill their own children. But there is no need of irrational gods to give rise to tragic situations.

Perhaps, however, it might be conceded that men may have regrets in these situations; it might even be conceded that a fully admirable moral agent would, on occasion, have such regrets; nevertheless (it may be said) this is not to be connected directly with the structure of the moral conflict. The man may have regrets because he has had to do something distressing or appalling or which in some way goes against the grain, but this is not the same as having regrets because he thinks that he has done something that he ought not to have done, or not done something that he ought to have done—and it is only the latter that can be relevant to the interpretation of the moral conflict. This point might be put, in terms which I hope will be recognisable, by saying that regrets may be experienced in terms of purely *natural* motivations, and these are not to be confused, whether by the theorist or by a rational moral agent, with *moral* motivations, i.e., motivations that spring from thinking that a certain course of action is one that one ought to take.

There are three things I should like to say about this point. First, if it does concede that a fully admirable moral agent might be expected to experience such regrets on occasion, then it concedes that the notion of such an agent involves his having certain natural motivations as well as moral ones. This concession is surely correct, but it is unclear that it is allowed for in many ethical theories. Apart from this, however, there are two other points that go further. The sharp distinction that this argument demands between these natural and moral motivations is unrealistic. Are we really to think that if a man (*a*) thinks that he ought not to cause needless suffering and (*b*) is distressed by the fact or prospect of his causing needless suffering, then (*a*) and (*b*) are just two separate facts about him? Surely (*b*) can be one expression of (*a*), and (*a*) one root of (*b*)? And there are other possible connexions between (*a*) and (*b*) besides these. If such connexions are admitted, then it may well appear absurdly unrealistic to try to prise apart a man's feeling regrets about what he has done and his thinking that what he has done is something that he ought not to have done, or constituted a failure to do what he ought to have done. This is not, of course, to say that it is impossible for moral thoughts of this type, and emotional reactions or motivations of this type, to occur without each other; this is clearly possible. But it does not follow from this that if a man does both have moral thoughts about a course of action and certain feelings of these types related to it, then these items have to be clearly and distinctly separable one from another. If a man in general thinks that he ought not to do a certain thing, and is distressed by the thought of doing that thing; then if he does it, and is distressed at what he has done, this distress will probably have the shape of his thinking that in doing that thing, he has done something that he ought not to have done.

The second point of criticism here is that even if the sharp distinction between natural and moral motivations were granted, it would not, in the matter of regrets, cover all the cases. It will have even the appearance of explaining the cases only where the man can be thought to have a ground of regret or distress independently of his moral opinions about the situation. Thus if he has caused pain, in the course of acting (as he sincerely supposes) for the best, it might be said that any regret or distress he feels about having caused the pain is independent of his views of whether in doing this, he did something that he ought not to have done: he is just naturally distressed by the thought of having caused pain. I have already said that I find this account unrealistic, even for such cases. But there are other cases in which it could not possibly be sustained. A man may, for instance, feel regret because he has broken a promise in the course of acting (as he sincerely supposes) for the best; and his regret at having broken the promise must surely arise *via* a moral thought. Here we seem just to get back to the claim that such regret in such circumstances would be irrational, and to the previous answer that if this claim is intended pejoratively, it will not stand up. A tendency to feel regrets, particularly creative regrets, at having broken a promise even in the course of acting for the best might well be considered a reassuring sign that an agent took his promises seriously. At this point, the objector might say that he still thinks the regrets irrational, but that he does not intend 'irrational' pejoratively: we must rather admit that an admirable moral agent is one who on occasion is irrational. This, of course, is a new position: it may well be correct.

6

It seems to me a fundamental criticism of many ethical theories that their accounts of moral conflict and its resolution do not do justice to the facts of regret and related considerations: basically because they eliminate from the scene the *ought* that is not acted upon. A structure appropriate to conflicts of belief is projected on to the moral case; one by which the conflict is basically adventitious, and a resolution of it disembarrasses one of a mistaken view which for a while confused the situation. Such an approach must be inherent in purely cognitive accounts of the matter; since it is just a question of which of the conflicting *ought* statements is true, and they cannot both be true, to decide correctly for one of them must be to be rid of error with respect to the other—an occasion, if for any feelings, then for such feelings as relief (at escaping mistake), self-congratulation (for having got the right answer), or possibly self-criticism (for having so nearly been misled). Ross—whom unfairly I shall mention without discussing in detail—makes a valiant attempt to get nearer to the facts than this, with his doctrine that the *prima facie* obligations are not just *seeming* obligations, but more in the nature of a claim, which can generate residual obligations if not fulfilled.[6] But it remains obscure how all this is supposed to be so within the general structure of his theory; a claim, on these views, must surely be a claim for consideration as the only thing that matters, a duty, and if a course of action has failed to make good this claim in a situation of conflict, how can it maintain in that situation some residual influence on my moral thought?

A related inadequacy on this issue emerges also, I think, in certain prescriptivist theories. Hare, for instance, holds that when I encounter a situation of conflict, what I have to do is modify one or both of the moral principles that I hold, which, in conjunction with the facts of the case, generated the conflict. The view has at least the merit of not representing the conflict as entirely adventitious, a mere misfortune that befalls my moral faculties. But the picture that it offers still seems inadequate to one's view of the situation *ex post facto*. It explains the origin of the conflict as my having come to the situation insufficiently prepared, as it were, because I had too simple a set of moral principles; and it pictures me as emerging from the situation better prepared, since I have now modified them—I can face a recurrence of the same situation without qualms, since next time it will not present me with a conflict. This is inadequate on two counts. First, the only focus that it provides for retrospective regret is that I arrived unprepared, and not that I did not do the thing rejected in the eventual choice. Second, there must surely be something wrong with the consequence that, granted I do not go back on the choice I make on this occasion, no similar situation later can possibly present me with a conflict. This may be a not unsuitable description of *some* cases, since one thing I may learn from such experiences is that some moral principle or view that I held was too naive or *simpliste*. But even among lessons, this is not the only one that may be learned. I may rather learn that I ought not to get into situations of this kind—and this lesson seems to imply very much the opposite of the previous one, since my reason for avoiding such situations in the future is that I have learned that in them both *ought's do* apply. In extreme cases, again, it may be that there is no lesson to be learned at all, at least of this practical kind.

7

So far I have been largely looking at moral conflict in itself; but this last point has brought us to the question of avoiding moral conflict, and this is something that I should like to discuss a little further. It involves, once more, but in a different aspect, the relations between conflict and rationality. Here the comparison with beliefs and desires is once more relevant. In the case of beliefs, we have already seen how it follows from the nature of beliefs that a conflict presents a problem, since conflicting beliefs cannot both be true, and the aim of beliefs is to be true. A rational man in this respect is one who (no doubt among other things) so conducts himself that this aim is likely to be realised. In the case of desires, again, there is something in the nature of desires that explains why a conflict essentially presents a problem: desires, obviously enough, aim at satisfaction, and conflicting desires cannot both be satisfied. Corresponding to this there will be a notion of practical rationality, by which a man will be rational who (no doubt among other things) takes thought to prevent the frustration of his desires. There are, however, two sides to such a policy: there is a question, not only of how he satisfies the desires he has, but of what desires he has. There is such a thing as abandoning or discouraging a desire which in conjunction with others leads to frustration, and this a rational man will sometimes do. This aspect of practical rationality can be exaggerated, as in certain moralities (some well known in antiquity) which avoid frustration of desire by reducing desire to a minimum: this can lead to the result that, in pursuit of a coherent life, a man misses out on the more elementary requirement of having a life at all. That this is the type of criticism appropriate to this activity is important: it illustrates the sense in which a man's policy for organising his desires is *pro tanto* up to him, even though some ways a man may take of doing this constitute a disservice to himself, or may be seen as, in some rather deeper way, unadmirable.

There are partial parallels to these points in the sphere of belief. I said just now that a rational man in this sphere was (at least) one who pursued as effectively as possible truth in his beliefs. This condition, in the limit, could be satisfied by a man whose sole aim was to avoid falsity in his beliefs, and this aim he might pursue by avoiding, so far as possible, belief: by cultivating scepticism, or ignorance (in the sense of never having heard of various issues), and of the second of these, at least, one appropriate criticism might be similar to one in the case of desires, a suggestion of self-impoverishment. There are many other considerations relevant here, of course; but a central point for our present purpose does stand, that from the fact that given truths or a given subject-matter exist, it does not follow that a given man ought to have beliefs about them: though it does follow that if he is to have beliefs about them, there are some beliefs rather than others that he ought to have.

In relation to these points, I think that morality emerges as different from both belief and desire. It is not an option in the moral case that possible conflict should be avoided by way of scepticism, ignorance, or the pursuit of *ataraxia*—in general, by indifference. The notion of a moral claim is of something that I may not ignore: hence it is not up to me to give myself a life free from conflict by withdrawing my interest from such claims.

It is important here to distinguish two different questions, one moral and one logical. On the one hand, there is the question whether extensive moral indifference is morally deplorable, and this is clearly a moral question, and indeed one on which two views are possible: *pas trôp de zèle* could be a moral maxim. That attitude, however, does not involve saying that there are moral claims, but it is often sensible to ignore them; it rather says that there are fewer moral claims than people tend to suppose. Disagreement with this attitude will be moral disagreement, and will involve, among other things, affirming some of the moral claims which the attitude denies. The logical question, on the other hand, is whether the relation of moral indifference and moral conflict is the same as that of desire-indifference and desire-conflict, or, again, belief-indifference and belief-conflict. The answer is clearly 'no'. After experience of these latter sorts of conflict, a man may try to cultivate the appropriate form of indifference while denying nothing about the nature of those conflicts as, at the time, he took them to be. He knows them to have been conflicts in believing the truth or pursuing what he wanted, and, knowing this, he tries to cut down his commitment to believing or desiring things. This may be sad or even dotty, but it is not actually inconsistent. A man who retreats from moral conflict to moral indifference, however, cannot at the same time admit that those conflicts were what, at the time, he took them to be, viz., conflicts of moral claims, since to admit that there exist moral claims in situations of that sort is incompatible with moral indifference towards those situations.

The avoidance of moral conflict, then, emerges in two ways as something for which one is not merely free to devise a policy. A moral observer cannot regard another agent as free to restructure his moral outlook so as to withdraw moral involvement from the situations that produce conflict; and the agent himself cannot try such a policy, either, so long as he regards the conflicts he has experienced as conflicts with a genuine moral basis. Putting this together with other points that I have tried to make earlier in this paper, I reach the conclusion that a moral conflict shares with a conflict of desires, but not with a conflict of beliefs, the feature that to end it in decision is not necessarily to eliminate one of the conflicting items: the item that was not acted upon may, for instance, persist as regret, which may (though it does not always) receive some constructive expression. Moral conflicts do not share with conflicts of desire (nor yet with conflicts of belief) the feature that there is a general freedom to adopt a policy to try to eliminate their occurrence.

It may well be, then, that moral conflicts are in two different senses ineliminable. In a particular case, it may be that neither of the *ought's* is eliminable. Further, the tendency of such conflicts to occur may itself be ineliminable, since, first, the agent cannot feel himself free to reconstruct his moral thought in a policy to eliminate them; and, second, while there are *some* cases in which the situation was his own fault, and the correct conclusion for him to draw was that he ought not to get into situations of that type, it cannot be believed that all genuine conflict situations are of that type.

Moral conflicts are neither systematically avoidable, nor all soluble without remainder.

8

If we accept these conclusions, what consequences follow for the logic of moral thought? How, in particular, is moral conflict related to logical inconsistency? What I have to say is less satisfactory than I should like; but I hope that it may help a little.

We are concerned with conflicts that have a contingent basis, with conflict via the facts. We distinguished earlier two types of case: that in which it seems that I ought to do a and that I ought to do b, but I cannot do both; and that in which it seems that I ought to do c in respect of some considerations, and ought not to do c in respect of others. To elicit something that looks like logical inconsistency here obviously requires in the first sort of case extra prem-isses, while extra premisses are at least not obviously required in the second case. In the second case, the two conclusions 'I ought to do c' and 'I ought not to do c' already wear the form of logical incon-sistency. In the first case, the pair 'I ought to do a' and 'I ought to do b' do not wear it at all. This is not surprising, since the conflict arises not from these two alone, but from these together with the statement that I cannot do both a and b. How do these three together acquire the form of logical inconsistency? The most natural account is that which invokes two further premisses or rules: that *ought* implies *can,* and that 'I ought to do a' and 'I ought to do b' together imply 'I ought to do a and b' (which I shall call the *agglom-eration principle*). Using these, the conflict can be represented in the following form:

(i) I ought to do a

(ii) I ought to do b

(iii) I cannot do a and b.

From (i) and (ii), by agglomeration

(iv) I ought to do a and b;

from (iii) by '*ought* implies *can*' used contrapositively,

(v) It is not the case that I ought to do a and b.

This produces a contradiction; and since one limb of it (v), has been proved by a valid inference from an undisputed premiss, we accept this limb, and then use the agglomeration principle contrapositively to unseat one or other of (i) and (ii).

This formulation does not, of course, produce an inconsistency of the *ought–ought not* type, but of the *ought–not ought* type, i.e., a genuine *contradiction*. It might be suggested, however, that there is a way in which we could, and perhaps should, reduce cases of this first type to the *ought–ought not* kind, i.e., to the pattern of the sec-ond type of case. We might say that 'I ought to do b', together with the empirical statement that doing a excludes doing b, jointly yield the conclusion that I ought to do something which, if I do a, I shall not do; hence that I ought to refrain from doing a; hence that I ought not to do a. This, with the original statement that I ought to do a, produces the *ought–ought not* form of inconsistency. A similar inference can also be used, of course, to establish that I ought not to do b, a conclusion which can be similarly joined to the original statement that I ought to do *b*. To explore this suggestion thoroughly would involve an extensive journey on the troubled waters of deontic logic; but I think that there are two considerations that suggest that it is not to be preferred to the formulation that I advanced earlier. The first is that the principle on which it rests looks less than compelling in purely logical terms: it involves the substitution of extensional equivalences in a modal context, and while this might possibly fare better with *ought* than it does elsewhere, it would be rash to embrace it straight off. Second, it suffers from much the same defect as was noticed much earlier with a parallel situation with conflicts of desires: it conceals the real roots of the conflict. The fomulation with '*ought* implies *can*' does not do this, and offers a more realistic picture of how the situation is.

Indeed, so far from trying to assimilate the first type of case to the second, I am now going to suggest that it will be better to assimilate the second to the first, as now interpreted. For while 'I ought to do *c*' and 'I ought not to do *c*' do indeed wear the form of logical inconsistency, the blank occurrence of this form itself depends to some extent on our having left out the real roots of the conflict— the considerations or aspects that lead to the conflicting judgements. Because of this, it conceals the element that is in common between the two types of case: that in both, the conflict arises from a contingent impossibility. To take Agamemnon's case as example, the basic *ought's* that apply to the situation are presumably that he ought to discharge his responsibilities as a commander, further the expedition, and so forth; and that he ought not to kill his daughter. Between these two there is no inherent inconsistency. The conflict comes, once more, in the step to action: that as things are, there is no way of doing the first without doing the second. This should encourage us, I think, to recast it all in a more artificial, but perhaps more illuminating way, and say that here again there is a double *ought:* the first, to further the expedition, the second, to refrain from the killing; and that as things are he cannot discharge both.

Seen in this way, it seems that the main weight of the problem descends on to '*ought* implies *can*' and its application to these cases; and from now on I shall consider both types together in this light. Now much could be said about '*ought* implies *can*', which is not a totally luminous principle, but I shall forgo any general discussion of it. I shall accept, in fact, one of its main applications to this problem, namely that from the fact that I cannot do both *a* and *b* it follows contrapositively that it is not the case that I ought to do both *a* and *b*. This is surely sound, but it does not dispose of the logical problems: for no agent, conscious of the situation of conflict, in fact thinks that he ought to do *both* of the things. What he thinks is that he ought to do *each* of them; and this is properly paralleled at the level of 'can' by the fact that while he cannot do both of the things, it is true of each of the things, taken separately, that he can do it.

If we want to emphasise the distinction between 'each' and 'both' here, we shall have to look again at the principle of agglomeration, since it is this that leads us from 'each' to 'both'. Now there are certainly many characterisations of actions in the general field of evaluation for which agglomeration does not hold, and for which what holds of each action taken separately does not hold for both taken together: thus it may be *desirable,* or *advisable,* or *sensible,* or *prudent,* to do *a,* and again desirable or advisable, etc., to do *b,* but not desirable, etc., to do both *a* and *b.* The same holds, obviously enough, for what a man wants; thus marrying Susan and marrying Joan may be things each of which Tom wants to do, but he certainly does not want to do both. Now the mere existence of such cases is obviously not enough to persuade anyone to give up agglomeration for *ought,* since he might reasonably argue that *ought* is different in this respect; though it is worth noting that anyone who is disposed to say that the sorts of characterisations of actions that I just mentioned are evaluative *because they entail 'ought'-statements* will be under some pressure to reconsider the agglomerative properties of *ought.* I do not want to claim, however, that I have some knock-down disproof of the agglomeration principle; I want to claim only that it is not a self-evident datum of the logic of *ought,* and that if a more realistic picture of moral thought emerges from abandoning it, we should have no qualms in abandoning it. We can in fact see the problem the other way round: the very fact that there can be two things, each of which I ought to do and each of which I can do, but of which I cannot do both, shows the weakness of the agglomeration principle.

Let us then try suspending the agglomeration principle, and see what results follow for the logical reconstruction of moral conflict. It is not immediately clear how '*ought* implies *can*' will now bear on the issue. On the one hand, we have the statement that I cannot do both *a* and *b,* which indeed disproves that I ought to do both *a* and *b,* but this is uninteresting: the statement it disproves is one that I am not disposed to make in its own right, and which does not follow (on the present assumptions) from those that I am disposed to make. On the other hand, we have the two *ought* statements and their associated 'can' statements, each of which, taken separately, I can assert. But this is not enough for the conflict, which precisely depends on the fact that I cannot go on taking the two sets separately. What we need here, to test the effect of '*ought* implies *can*', is a way of applying to each side the fact that I cannot satisfy both sides. Language provides such a way very readily, in a form which is in fact the most natural to use in such deliberations:

(i) If I do *b*, I will not be able to do *a*;
(ii) If I do *a*, I will not be able to do *b*.

Now (i) and (ii) appear to be genuine conditional statements; with suitable adjustment of tenses, they admit both of contraposition and of use in *modus ponens*. They are thus not like the curious nonconditional cases discussed by Austin.[7]

Consider now two apparently valid applications of '*ought* implies *can*':

(iii) If I will not be able to do *a*, it will not be the case that I ought to do *a*;
(iv) If I will not be able to do *b*, it will not be the case that I ought to do *b*.

Join (iii) and (iv) to (i) and (ii) respectively, and one reaches by transitivity:

(v) If I do *b*, it will not be the case that I ought to do *a*;
(vi) If I do *a*, it will not be the case that I ought to do *b*.

At first glance (v) and (vi) appear to offer a very surprising and reassuring result: that whichever of *a* and *b* I do, I shall get off the moral hook with respect to the other. This must surely be too good to be true; and suspicion that this is so must turn to certainty when one considers that the previous argument would apply just as well if the conflict between *a* and *b* were not a conflict between two *ought*'s at all, but, say, a conflict between an *ought* and some gross inclination; the argument depends solely on the fact that *a* and *b* are empirically incompatible. This shows that the reassuring interpretation of (v) and (vi) must be wrong. There is a correct interpretation, which reveals (v) and (vi) as saying something true but less interesting: (taking (v) as example), that if I do *b*, it will then not be correct to say that I ought (then) to do *a*. And this is correct, since *a* will *then* not be a course of action open to me. It does not follow from this

that I cannot correctly say then that *I ought to have done a;* nor yet
that I was wrong in thinking earlier that *a* was something I ought to
do. It seems, then, that if we waive the agglomeration principle, and
just consider a natural way of applying to each course of action the
consideration that I cannot do both it and the other one, we do not
get an application of *ought* implies *can* that necessarily cancels out
one or other of the original *ought's* regarded retrospectively. And
this seems to me what we should want.

As I have tried to argue throughout, it is surely falsifying of
moral thought to represent its logic as demanding that in a conflict
situation one of the conflicting *ought's* must be totally rejected. One
must, certainly, be rejected in the sense that not both can be acted
upon; and this gives a (fairly weak) sense to saying that they are
incompatible. But this does not mean they do not both (actually)
apply to the situation; or that I was in some way mistaken in thinking that these were both things that I ought to do. I may continue
to think this retrospectively, and hence have regrets; and I may even
do this when I have found some moral reason for acting on one in
preference to the other. For while there are some cases in which
finding a moral reason for preference *does* cancel one of the *ought's,*
this is not always so. I may use some emergency provision, of a utilitarian kind for example, which deals with the conflict of choice,
and gives me a way of 'acting for the best'; but this is not the same
as to revise or reconsider the reasons for the original *ought's,* nor
does it provide me with the reflexion 'If I had thought of that in the
first place, there need have been no conflict.' It seems to me impossible, then, to rest content with a logical picture which makes it a
necessary consequence of conflict that one *ought* must be totally
rejected in the sense that one becomes convinced that it did not
actually apply. The condition of moving away from such a picture
appears to be, at least within the limits of argument imposed by my
rather crude use of *ought* implies *can,* the rejection of the agglomeration principle.

I have left until last what may seem to some the most obvious
objection to my general line of argument. I have to act in the conflict; I can choose one course rather than the other; I can think about
which to choose. In thinking about this, or asking another's advice
on it, the question I may characteristically ask is 'what ought I to
do?' The answer to this question, from myself or another, cannot be
'both', but must rather be (for instance) 'I (or you) ought to do *a.*'
This (it will be said) just shows that to choose in a moral conflict,

or at least to choose as a result of such deliberation, is to give up one of the *ought's* completely, to arrive at the conclusion that it does not apply; and that it cannot be, as I have been arguing that it may be, to decide not to act on it, while agreeing that it applies.

This objection rests squarely on identifying the *ought* that occurs in statements of moral principle, and in the sorts of moral judgements about particular situations that we have been considering, with the *ought* that occurs in the deliberative question 'what ought I to do?' and in answers to this question, given by myself or another. I think it can be shown that this identification is a mistake, and on grounds independent of the immediate issue. For suppose I am in a situation in which I think that I ought (morally) to do *a,* and would merely very much like to do *b,* and cannot do both. Here, too, I can presumably ask the deliberative question 'what ought I to do?' and get an answer to it. If this question meant 'Of which course of action is it the case that I ought (morally) to do it?', the answer is so patent that the question could not be worth asking: indeed, it would not be a deliberative question at all. But the deliberative question can be worth asking, and I can, moreover, intelligibly arrive at a decision, or receive advice, in answer to it which is offensive to morality. To identify the two *ought's* in this sort of case commits one to the necessary supremacy of the moral; it is not surprising if theories that tend to assimilate the two end up with the Socratic paradox. Indeed, one is led on this thesis not only to the supremacy, but to the ubiquity, of the moral; since the deliberative question can be asked and answered, presumably, in a situation where neither course of action involves orginally a moral *ought.*

An answer to the deliberative question, by myself or another, can of course be supported by moral reasons, as by other sorts; but its role as a deliberative *ought* remains the same, and this role is not tied to morality. This remains so even in the case in which both the candidates for action that I am considering involve moral *ought's.* This, if not already clear, is revealed by the following possibility. I think that I ought to do *a* and that I ought to do *b,* and I ask of two friends 'what ought I to do?'. One says 'You ought to do *a',* and gives such-and-such moral reasons. The other says 'You ought to do neither: you ought to go to the pictures and give morality a rest.' The sense of *ought* in these two answers is the same: they are both answers to the unambiguous question that I asked.

All this makes clear, I think, that if I am confronted with two conflicting *ought's,* and the answer to the deliberative question by myself or another *coincides* with one of the original *ought's,* it does not represent a mere *iteration* of it. The decision or advice is decision or advice to act on that one; not a reassertion of that one with an implicit denial of the other. This distinction may also clear up what may seem troubling on my approach, that a man who has had a moral conflict, has acted (as he supposes) for the best, yet has the sorts of regrets that I have discussed about the rejected course of action, would not most naturally express himself with respect to that course of action by saying 'I ought to have done the other.' This is because the standard function of such an expression in this sort of situation would be to suggest a deliberative mistake, and to imply that if he had the decision over again he would make it differently. That he cannot most naturally say this in the imagined case does not mean that he cannot think of the rejected action as something which, in a different sense, he ought to have done; that is to say, as something of which he was not wrong at the time in thinking that he ought to do it.

In fact, of course, it is not even true that *the* deliberative question is 'what ought I to do?'. It may well be, for instance, 'what am I to do?'; and that question, and the answers to it—such as 'do *a*', or 'if I were you, I should . . . '—do not even make it look as though decision or advice to act on one of the *ought's* in a moral conflict necessarily involves deciding that the other one had no application.

Notes

1. I shall in the rest of this paper generally use the phrase 'conflict of beliefs' for the situation in which a man has become conscious that his beliefs conflict.

2. For a discussion of a similar notion, see A. Kenny, *Action, Emotion and Will* (London: Routledge & Kegan Paul, 1963), chapter 5.

3. Plato, incidentally, seems to have thought that all conflicts that did not involve a moral or similar motivation had a contingent basis. The argument of *Republic* IV which issues in the doctrine of the divisions of the soul bases the distinction between the rational and epithymetic parts on conflicts of desire and aversion directed towards the same object in the same respects. But not all conflicts establish different parts of the soul: the epithymetic part can be in conflict with itself. These latter conflicts, therefore, cannot be of desires directed towards the same object in the same respects; that is to say, purely epithymetic conflicts have a contingent basis.

4. Here and elsewhere in this chapter Williams is referring to Hare's early books, *The Language of Morals* and *Freedom and Reason,* rather than *Moral Thinking,* from which chapter 11 in this volume is taken.—C.W.G.

5. Aeschylus, *Agamemnon* 217.

6. Cf., *The Foundations of Ethics* (Oxford: The Clarendon Press, 1938), pp. 84 *seq.* The passage is full of signs of unease; he uses, for instance, the unhappy expression 'the most right of the acts open to us', a strong indication that he is trying to have it both ways at once. Most of the difficulties, too, are wrapped up in the multiply ambiguous phrase 'laws stating the tendencies of actions to be obligatory in virtue of this characteristic or of that' (p. 86).

7. *Ifs and Cans,* reprinted in his *Philosophical Papers* (Oxford: The Clarendon Press, 1961).

7

Values and the Heart's Command

Bas C. van Fraassen

Since its inception deontic logic has been plagued by a series of paradoxes. In an earlier paper I worked out a solution to those paradoxes, which required an extension and reinterpretation of, but no deviation from, what have come to be regarded in deontic logic as "orthodox" principles.[1] Since then I have come to the opinion that those principles themselves reflect a serious flaw in the philosophical foundations of deontic logic. In this paper I shall explain my misgivings and attempt to substantiate them; technical results in deontic logic suggested by this critique will be indicated briefly.

I. The Axiological Thesis

If, when instructed concerning what ought to be, I ask for reasons, the answer may be in terms of duties, obligations, rights, ideals (of justice, of goodness, of fairness), or values (moral, aesthetic, reli-

From *The Journal of Philosophy* 70 (1973): 5–19. Reprinted with the permission of *The Journal of Philosophy* and the author.

gious). I may not have exhausted the possibilities. With inevitable simplicity, philosophers have divided the study of these reasons into two broad fields: *deontology* (theory of obligations; from the Greek δεον, *that which is binding, needful, proper*) and *axiology* (theory of values; from αξιος, *worth*, as in *is worth more than*). The former deals with what ought to be because it is required by one's station and its duties, by the web of obligations and commitments the past has spun. The latter deals with what ought to be because its being so would be good, or at least better than its alternatives.

With equal inevitability, there have been attempts to pare the two to one. The axiological thesis is stated succinctly by Moore:

> . . . to assert that a certain line of conduct is, at a given time, absolutely right or obligatory, is obviously to assert that more good or less evil will exist in the world, if it is adopted, than if anything else be done instead.[2]

And this thesis can claim ample support from actual usage and actual patterns of admonition. Thus St. Paul, that most Apollonian of Christians, phrases his imperatives in axiological terms: "It is good for a man not to touch a woman," he says, "But if they cannot contain, let them marry: for it is better to marry than to burn" (I Cor. 7).

There is a common argument that such a reduction to axiology is mistaken.[3] For, it is argued, there are many cases in which we agree that although it would be better for something to be, there is no obligation to bring it about. This, the problem of supererogation, raises havoc with the thesis as stated by Moore. But perhaps a less crude formulation will avoid the problem. Let us agree that some, and only some, true statements of the form "It ought to be the case that *A*" are true because of the existence of relevant obligations. We can then insist that, still, *all* of them are true because (and exactly because) it would be better if what they prescribe were the case. This means agreeing to the assertion that, really, people *ought* to act above and beyond the call of duty, when that is for the best, although they cannot (and, perhaps, ought not) be held to account, in some sense, if they do not.

In this attenuated form, the axiological thesis amounts to: there is some scale of values whereby what ought to be is exactly what is better on the whole. It is a thesis not concerning the relation between obligations and values, but between values and what ought to be; nor does it deny any relation between obligations and what ought to be. (With a bit more charity, we can even ascribe the thesis to Moore in its attenuated form; he would not have been the last to confuse the issues by neglecting the distinction between what is obligatory, and hence ought to be, and what ought to be for some reason or other.)

II. Implications of the Axiological Thesis

It should be noted that the very use of such phrases as 'more good', 'less evil', 'better', and 'best', presupposes that there is a *single* scale of values to which all moral judgments defer. For if neither of two courses of action is any better than the other, then it follows that they are equally good. (It would be quite different if Moore and St. Paul had used such terms as 'more good of some kind' or 'better in some respect'; for 'better in some respect' is compatible with 'worse (all told)'. And it is possible that each of two courses of action be no better than the other in some respect, and also that there be no respect in which they are equally good or bad.)

So, from this point of view, whether an ought-statement is true depends on two factors: the set of alternative possibilities we are evaluating, and the scale of values by which we rate them.

If the function of the ought-statement is to *counsel,* the alternatives considered are the possible outcomes of our action. If the function is to *judge,* the alternatives are the actual state of affairs and those which might have been actual if we had acted differently. For any sentence A, let $H(A)$ be those alternatives which make A true. Then we say:

> "It ought to be the case that A" is true exactly if some value attaching to some outcome in $H(A)$ is higher than any attaching to any outcome in $H(\text{not } A)$.

Intuitively phrased: we ought to opt for the realization of the highest possible values, and, more generally, for any state of affairs that is a necessary condition for the realization of the highest attainable values. (To obviate an obvious objection: when a possible outcome is evaluated, its relative likelihood should be given due attention.)

This is easily generalized to conditional ought-statements such as "The poor ought to be succored, if there are any," written canonically as "It ought to be that A, if B." For this we note that if B be *given* as true and an inescapable part of the context, then in judging or counseling one must ignore all those logically possible states in which B is not true. So in the above truth-definition we replace $H(\_\_\_)$ by $H_B(\_\_\_)$, which is the same as $H(B\&\_\_\_)$. Intuitively, it ought to be that A, if B, exactly if $(B \text{ and } A)$ is, on the whole, better than $(B \text{ and not } A)$.

The logic thus founded has a quite traditional look, counting as tradition the still very short history of the subject (see the paper mentioned in note 1).

III. Moral Conflicts

If the axiological thesis, even in its attenuated form, is accepted, then certain tenable ethical positions are ruled out. From this (which I shall try to substantiate) I conclude that the axiological thesis is itself an ethical doctrine, not a thesis of metaethics. (And if that is so, deontic logic should not be founded upon it, although a logic so founded might be interesting as a special subject from a metaethical point of view.) More important, perhaps, is this corollary: notwithstanding any slogans about how fulfilling obligations, or having good intentions, or living by ideals, is something good (or good on the whole), not every ethical doctrine differs from, say, utilitarianism only in its choice of values.

To demonstrate this, I must show what is ruled out by the axiological thesis. And what is ruled out is exactly this: that there should ever be an unresolvable ethical conflict. By this I mean a conflict between what ought to be for one reason and what ought to be for another reason, which cannot be resolved in terms of one reason overriding another, or one law or authority or value being higher than another. Sir David Ross thought that it was exactly the business of the moral philosopher to show how such conflicts are to be resolved, how conflicts between *prima facie* duties are, after all, illusory upon proper understanding. The axiological thesis requires this. For suppose that A and B are incompatible. Then if it ought to be the case that A, higher values attach to some outcomes satisfying A than to any that satisfy *not A*. But, because of the assumed incompatibility, all outcomes that satisfy B satisfy *not A*. Hence it is better to opt for A than for B. So, whenever A and B are mutually incompatible, it cannot be that both ought to be the case—either we ought to opt for A, or we ought to opt for B, or the matter is indifferent (morally indifferent, that is).

Is this a substantive ethical assumption, or is it nigh tautological? I shall present the case for both sides: for its denial and for its assertion, and hope to show that it is indeed a thesis which, although phrased in metaethical terminology, actually concerns the kind of fact of moral life on which ethical theories founder.

IV. Dirty Hands

It will occasion no surprise if some primitive ethic places its adherents in terrible quandaries, resolvable only by riding roughshod over one or another of its demands. If Alfonso van Worden, whose whole conception of moral action is governed by that of the point of honor,[4] finds himself in dire moral predicaments, we shan't draw theoretical conclusions. These are the perils of moral barbarism. But turning from barbarians to Greeks, we find Orestes torn between two clear moral commandments, badgered into one course of action by a god's threats and a sister's flattery, and the conflict finally resolved not by a decisive moral argument but by a political settlement between gods and Furies. And closer to us is Nora's reply to Helmer's assertion that her most sacred duty is to her husband and children: "I have another duty, just as sacred. . . . My duty to myself."[5] If two duties, equally sacred, conflict, an exercise of the will can settle the conflict, but not a calculation of values.

Few moral theorists seem to have accepted the appearance that there are real conflicts, as opposed to merely *prima facie* conflicts. However, there is a sense in which they occur in Bradley's ethical scheme [see chapter 3 of this volume]. (The qualification is that for Bradley all the ethical considerations apply only to Appearance and not to Reality; but the final resolution of all contradictions in the Absolute is not the kind of resolution of all ethical conflicts that is being denied here.)

In Bradley's *Ethical Studies,* his own ethical theory is set forth dialectically, in the typical idealist thesis-antithesis-synthesis pattern. We need look only at the last two stages. The penultimate stage is the morality of "My Station and Its Duties." Each man has, through circumstances and through choice, a certain place in the social structure. This place is largely characterized through the set of duties and rights that accrue to the occupier of that place. We may here construe the social structure widely, so that all relations to others (including blood relations and relations through commitments, moral debts, and promises) enter into the definition of one's station therein. And while certain aspects of this station are due to the circumstances of one's birth (consider the obligation one has qua son, qua citizen), there is no implication here of a "rigid" socio-economic structure. To live morally is to live as required by one's station. The intuitive evidence for this is considerable: it is certainly *prima facie* immoral to break one's promises, ignore filial duties, renege on commitments, and so forth.

But, as Bradley argues, the morality of this penultimate stage (even if surrounded with safeguards concerning the presuppositions of obligation, so that war criminals cannot plead duty, for instance), is unacceptably shortsighted. There are also the demands of self-realization, the ideals which, we fancy, guide our progress from the state of nature to a life proper to those created only a little lower than the angels. These ideals include moral ideals: it ought to be, from the moral point of view, that man live in freedom, freedom from subjugation and freedom to pursue the ends of self-knowledge, personal sanctity, control over his mind and passions—or, if the catalogue should be less puritan, to pursue to its limits the possibility of human experience, and the understanding thereof. But though our souls cry to high heaven for perfection, no such perfection is instantiated in our actual communities, and the possibilities we contemplate are closed by force of circumstance. The ideal of a life in freedom and love creates its own categorical imperative, but our station's duties may require competition, aggression, and exploitation.

It may seem that, in such a situation, there is a resolution, a morally correct resolution, of the conflict, namely revolution. The intent of revolution is so to change the social and economic structure that it will no longer be incompatible with the ideals that the revolutionary sees everywhere violated. But the revolutionary places himself in a role subject to a conflict of exactly similar structure. During the process of revolution, the pursuit of many ideals, and the exercise of rights, is temporarily suspended by a force of circumstances more violent than the norm. The revolutionary himself proposes thus to sacrifice himself and if necessary his whole generation for the sake of the coming community that he serves. He chooses the second horn of the dilemma described by Bradley, not wrapping himself "in a virtue that is [his] own and not the world's" but accepting "through faith and through faith alone, [that] self-suppression issues in a higher self-realization."[6] The resolution is through will and through faith, not through moral argument alone.

Closer to our time, Sartre has maintained that no ethical system can resolve all moral dilemmas. (We must take him to be referring to actual ethical systems, not possible systems concocted by logicians.) As example he considers Christian morality in connection with a case of a French boy who, at the start of the German occupation, must choose between joining the Free French and seeing his aged mother through the coming ordeal. "Qui doit-on aimer comme son frère, le combattant ou la mère?"[7]

Sartre gives this thesis practical content by arguing that in any political context (taken in a wide sense) effective action presupposes the will to countenance violence.[8] Hoederer states the view even more strongly in *Les Mains Sales:* you cannot act effectively without dirtying your hands. Hugo, however, cannot accept this: he feels with equal strength the command not to kill and the command to work for the classless society; he cannot accept that one overrides the other. Neither the godless Communist nor the god-possessed capitalist would have his dilemma, but for Hugo the conclusion is that if you can't act effectively without dirty hands, then guilt is inevitable. In *Le Diable et le Bon Dieu,* a kind of sequel to the other play, Goetz comes to terms with the problem: he resolves by an act of will what he cannot resolve by moral argument—and without pretending that what he does is morally right or most right. The existentialist hero is the man who does not fail to act upon his ideals, *and* does not rationalize away his dirty hands.

V. Counterarguments

I have presented this particular view of the moral situation at some length, because, if it is tenable, then the axiological thesis is incorrect. (Note that I say "tenable," not "true"; *possibly,* a number of alternative stances with respect to morality are tenable, and, *possibly,* the philosopher cannot guide the ultimate choice between competing moral views.) But there are a number of arguments designed to show this view untenable, and to these I turn now.

A reasonable man will strike a balance between the conflicting demands of moral duty and moral ideal. Perhaps so. Perhaps that is what 'reasonable' means. But it does not follow that there is a morally right balancing, in the sense in which there is a moral solution to the dilemma that occurs if a murder cannot be prevented without a lie. In that case one commandment clearly overrides another. But overriding is not a relationship that places all moral imperatives in a linear order.

Well, it may not *seem* to, there certainly are *prima facie* conflicts; but what reason is there to believe that careful moral argument will not always suffice to resolve them? That is a curious counterargument: if there is a conflict *prima facie,* the presumption should surely be that it is real, and the *onus probandi* is on those who say not. But I shall give a reason nevertheless. Law is somewhat like morality: through its historical development, a community develops a system of laws and a system of morals. There are often cases before the tribunal in which extant law rules ambiguously, inconsistently, or not at all. There is for such cases an intricate set of rules and procedures for adjusting the system of laws through creative interpretation by judges, precedent, legislative action, and plebiscites or elections. The similarity between law and morals suggests that there must similarly occur many cases in which our morality's guidance is ambiguous, inconsistent, or absent altogether. And for morality there is no institutionalized process of adjustment. (There is for those who can submit such questions to any agency they accept as infallible on questions of morals and faith, but whether that is right is itself a moral question.) And if the present moral conflicts disappear through a natural evolution over the generations, that is of little moment to the present, subject to present morality.

The two counterarguments given so far dealt with inclining reasons, but there are also counterarguments dealing with logic. It is asserted that "it ought to be the case that" implies "it is permitted (morally unobjectionable) that it be the case that," and that similarly "ought not" implies "not permitted." But then it follows that if it ought to be the case that A, then it is permitted that A, and hence it cannot be true that it ought not to be the case that A. Hence, A and *not A* can never both be such that they ought to be the case.

There are two, by now classical, principles of deontic logic that provide the impetus for this third counterargument. The first is that 'permitted' (or 'unobjectionable') is definable as "not ought not." I have no objection to the definition. The second is that "ought" implies "permitted" (so defined). But that is *equivalent* to the thesis that two oughts can never conflict. Symbolically:

$$O(A) \supset P(A)$$
$$O(A) \supset \sim O(\sim A)$$

$$\sim O(A) \lor \sim O(\sim A)$$
$$\sim [O(A)\ \&\ O(\sim A)]$$

(where I am using '$\sim$' for "not", '$\lor$' for "or", '&' for "and", 'O' for "ought", and 'P' for "permitted"). I can only conjecture that the original devisors of deontic axioms had a certain ethical bias; perhaps they were utilitarians, or accepted some other axiological creed.

The next logical counterargument also has two premises. The first is that "ought" implies "can." That is, no one can be subject to a moral obligation to do the impossible. The second is that logical consequences of what ought to be, ought to be. There may be some equivocation in that, but we can hardly expect to bring about what ought to be without bringing about its necessary consequences. So I won't object to either doctrine. But then, the argument goes on, suppose that I ought to see to it that A and also that B, where A and B are logically incompatible. Evidently I am then under a moral obligation to do the impossible, which is absurd.

Here I must object to the equivocation. Consider a (relatively) concrete example. The agent is subject to incompatible obligations due to his several allegiances to heaven and earth (sons and lovers, party and fatherland, choose what you will). He appears before the tribunal of heaven (respectively, of earth) and, pointing to his several allegiances, defends his shortcomings by the statement that he cannot be expected to do the impossible. Whereupon the heavenly judge points out, with irrefutable logic, that the agent is held guilty *not* of failing to do the impossible, but of failing to honor his allegiance to the cause of heaven. His defense before the earthly tribunal fares no better (recall Orestes' last conversation with his mother). The accused was equivocating between having a commitment to do A and also a commitment to do B, and having a commitment to do both A and B.

Here it might be interjected that someone required on the one hand to do A, and on the other to do something incompatible with A, is for all practical purposes required to do the impossible. But there is in fact a crucial difference. For to attain a state-of-affairs X, one must attain all necessary conditions of X. Now if X is impossible, then everything is a necessary condition of X (in the slipshod sense accepted everywhere outside the logic of relevant implication). So if one is required to do the impossible, one is required to do everything, and *all moral distinctions collapse*. But for the person in a moral quandary it is by no means true that all moral distinctions have collapsed—much as he might like to plead this.

The remaining counterarguments concern the notion of guilt. As a subject, this is practically *tabu* to moderns, among whom other four-letter Saxon words have become quite acceptable. (Perhaps it would help if we returned to the four-letter spelling.) I think I can at least clear myself of psychologism with the simple thesis that it is appropriate to feel guilt if and only if one is guilty, and that there is no overriding moral reason to indulge in any feeling, whether appropriate or not.

The first argument to consider here is that the problem is simply one of *misplaced* guilt. Not when he failed to honor one or other allegiance or commitment, but when he allowed himself to become subject to several allegiances or commitments that might lead to a quandary, was the agent guilty of an immoral action. However, this implies that we cannot act both reasonably and morally on probabilities as opposed to certainties. For it means that, no matter how good the consequences of accepting a commitment, its acceptance is morally wrong if there is the least possible chance that it could figure in a moral dilemma. Secondly, it seems to me that this counterargument assumes that obligations are typically incurred by conscious decisions. And this seems false.

The second and third counterarguments concerning guilt were added by Robert Stalnaker.[9] If we admit the possibility of a moral dilemma without resolution, then a judgment attributing guilt to the man in such a dilemma can be made without reference to his character or actions. But that is absurd: a moral judgment about a man can be justified only in terms of factors under his control.

This last assertion I consider to be itself a substantive ethical thesis, and not one that can be justified on the basis of what the terms mean, because it denies the doctrine of original sin (under at least one straightforward interpretation thereof). I quote, in free translation, the beginning of the Five Articles of the Synod of Dordrecht (1618–19): "Since all men have sinned in Adam, and have [thus] become so guilty as to deserve damnation and eternal death, God would have treated no one with injustice had He left the whole human race in sin and damnation and doomed it because of [this] sin."

The third and last such argument is perhaps the most basic. The proponent of the axiological thesis can make perfect sense of the situations described by his opponent. It is possible to be in a situation in which one course of action leads to the attainment of something of great value, and another course of action leads to the attainment of something quite different of equally great value. The choice will be agonizing, and if afterward one feels regret to the point of anguish, this is only natural, since something of great value has been lost. Thus the facts of the moral situation are not denied by this proponent; he just describes them differently.

The key term here is 'regret'. The case of Orestes or of Nora is counted as fundamentally the same as that of the philanthropist who regrets that he has but one fortune to give for mankind and has agonized over the choice between endowing the arts and furthering birth control (or as the case of the revolutionary who agonizes over the question where he should risk his life against oppression—in Bolivia or in Guatemala). But the cases are the same only if regret is the same as guilt, or if it is necessarily appropriate to feel guilt if and only if it is appropriate to feel regret. And that can tenably be denied.

VI. Imperatives and Truth

The metaethical argument that the view of morality examined was tenable proceeded by displaying the arguments and counterarguments of proponents and opponents of the view—hence by displaying ethical arguments. I shall conclude that the view constitutes a significant ethical position (whether correct or incorrect); let us return now to the airy heights of metaethics.

Ethical conflicts are possible; so sometimes there are two sound moral arguments, concluding respectively that A ought to be the case ($O(A)$) and that *not-A* ought to be the case ($O(\sim A)$). When we have arrived at two conclusions, we can conjoin them:

$$O(A) \ \& \ O(\sim A)$$

can be true. But "ought" implies "can" (or, at least, I see no reason to deny that it does); so

$$O(A \ \& \sim A)$$

cannot be true. Finally, logical consequences (i.e., necessary conditions) of what ought to be, ought to be; for a man ought not to put himself in a position in which it is impossible for him to do what he ought to do. Hence

If B can be inferred from A, then $O(B)$ can be inferred from $O(A)$.

is a sound rule.

But we already know that no sense can be made of the above if we construe what ought to be as what is better or for the best. To make sense of it I shall appeal to the idea of moral imperatives or commandments, taking a cue from Kant, the young Hegel, the moral intuitionists, Sellars, and Castañeda. However, I must confess a great deal of ignorance about the moral imperative and its logic. I shall sum up the little I think I know.

First, there are many conceivable imperatives, but only some are *in force*. The problem of the ontogenesis of moral imperatives is the problem of what brings imperatives in force. I assume their sources are legion: conscience, ideals, values, duties, commitments, and so on. The process must be complicated, because one imperative may be *overridden* by another (under given circumstances); this means presumably that one imperative's being in force may prevent another's being in force. In this way, direct orders may cancel standing orders, and circumstances may take away the authority under which a standing order is the reason that a certain imperative is in force.

When an imperative is in force, we evaluate possible outcomes and possible states of affairs with respect to it, by asking whether the imperative is *fulfilled* or *violated*. So for each imperative I there is a class of possible outcomes I^+ in which it is fulfilled. As an initial attempt to explain 'ought', I propose:

$O(A)$ is true exactly if, for some imperative I that is in force, I^+ is part of the set of possible outcomes in which A is true.

This means, less formally, that it ought to be that A exactly if some imperative in force would not be fulfilled if *not-A*.

Now, with this definition, $[O(A) \& O(\sim A)]$ could be true, namely if there were two conflicting imperatives in force. Secondly, $O(A \& \sim A)$ cannot be true if no *single* imperative is impossible to fulfill (taken by itself). And, if A implies B, then $O(A)$ implies $O(B)$; that also follows from the definition. A sound and complete logical system, for a language constructed with the above truth-condition for 'O' and subject to the restriction that any single imperative can possibly be fulfilled, has the following axiom schemes and rules:

A1.   Axiom schemata for propositional logic

A2. $\vdash \sim O(\sim A \& A)$

R1.   if $\vdash A$ and $\vdash A \supset B$, then $\vdash B$

R2.   if $\vdash A \supset B$, then $\vdash O(A) \supset O(B)$

Finally, under suitable assumptions about the ultimate resolution of all conflicts, the definition will agree with that of the axiologist.

But this scheme is too simpleminded, because it does not take account of two facts. The first is that imperatives are typically conditional; a venerable example is St. Paul's "Is any man called being circumcised? let him not become uncircumcised." (I Cor. 7). If we try to construe this as the categorical demand that some conditional be true, we may expect the Good Samaritan and its retinue to plague us again. (In fact, the Good Samaritan paradox was introduced by Bradley not against the axiologist but against the deontologist, citing Blake's ironic "Pity would be no more/If we did not make somebody poor.")[10]

So imperatives are themselves conditional, and I shall assume that a conditional imperative can be fulfilled *or* violated only if its condition is the case. If the Corinthian church numbered no Jews, they might subscribe to, accept, or take to heart what St. Paul said, but they could neither follow nor violate his injunction.

Suppose it is imperative that A, given B; and suppose that B is true (or rather, inevitable, since we are considering *outcomes* of courses of action). Does it follow that it is imperative that A? I would say no (though von Wright and Castañeda apparently disagree). My reason is that I take imperatives to have *presuppositions:* I take it that circumstances may remove the force of imperatives, or the authority or justification from which they derive their force. Also, under new circumstances, one imperative may come to be overridden by another. Hence I cannot accept that what is imperative if everything is possible, will be imperative if inevitable limits to action appear. There may be systematic relations governing this moral dynamics, but I can only profess ignorance of them.

The second scheme I propose therefore takes ought-statements to have a conditional form $O(A/B)$ ("it ought to be that A, given B"), and explains them in terms of conditional imperatives:

$O(A/B)$ is true exactly if there is some imperative I in force, which is itself conditional upon B, such that A is true in all the outcomes in which B is true and which fulfill I.

More briefly, $O(A/B)$ is true if I_B^+ H(B) is part of H(A), where by '$H(\_\_\_)$' we mean the set of outcomes in which $\_\_\_$ is true, and I_B is some imperative conditional upon B.

If a language is constructed with this truth-definition and if the assumption that any single imperative with possible antecedent can be in force only if it is possible that it be fulfilled, a sound and complete logical system has the following axiom schemes and rules:

AC1. Axiom schemata for propositional logic
AC2. $\vdash O(A/B) \supset O(A \ \& \ B/B)$
RC1. if $\vdash A$ and $\vdash A \supset B$ then $\vdash B$
RC2. if $\vdash A \supset B$ then $\vdash O(A/C) \supset O(B/C)$
RC3. if $\vdash B \equiv C$ then $\vdash O(A/B) \equiv O(A/C)$

With suitable additional assumptions guaranteeing the resolution of all conflicts, the correct logical system becomes equivalent to AC1–AC4 with RC1–RC4 of my "The Logic of Conditional Obligation."

VII. Evaluation by Imperatives

For simplicity, let us just consider unconditional imperatives and ought-statements, and see whether we have not oversimplified.

Though I have no good account to give of the idea that one imperative may *override* another, I have adopted the thesis that an imperative is not in force if it is overridden. Hence the relationship of overriding need not play a role in the account of the truth of ought-statements, since, there, only imperatives in force are relevant.

However, it seems natural to say that if one's choice is between fulfilling two imperatives (in force) and fulfilling only one of them, one ought to do the first. As example, Stalnaker proposed:

I. (a) Honor thy father or thy mother!
   (b) Honor not thy mother!
         Hence, thou shalt honor thy father.

It is important to see that the premises express imperatives, and are not ought-statements. For, since various incompatible imperatives may compete, the statement that one ought not to honor one's mother might be true because one ought not to honor either one's mother or one's father, owing to an imperative incompatible with (a).

Secondly, argument I is not valid if it is impossible to honor one's father, since "ought" implies "can."

However, even with these provisos it is clear that the account I gave in the preceding section does not do justice to the example. For if (a) and (b) are the only imperatives in force, then there is no imperative I in force such that $I^+ \subseteq H$ ("Thou honorest thy father").

Accordingly, I propose a revision of the truth-definition. Suppose that β is one of the possible alternatives we are considering. Let us say that the *score* of β is the class of imperatives in force that β fulfills. Then:

> O(A) is true if and only if there is a possible state of affairs β in H(A) whose score is not included in the score of any γ in H($\sim$A).

It can be seen immediately that the basic criteria are satisfied: $O(A)$ and $O(\sim A)$ can both be true; $O(A \ \& \sim A)$ cannot be true; and, if A implies B, then $O(A)$ implies $O(B)$. In addition, if I_1 and I_2 are the only imperatives in force and if there is a β that belongs to I_1^+ and also to I_2^+, then, if β is in H(A), $O(A)$ is true.

But can this happy circumstance be reflected in the logic of the ought-statements alone? Or can it be expressed only in a language in which we can talk directly about the imperatives as well? This is an important question because it is the question whether the inferential structure of the "ought" language game can be stated in so simple a manner that it can be grasped in and by itself. Intuitively, we want to say: there are simple cases, and in the simple cases the axiologist's logic is substantially correct even if it is not in general— but can we state precisely when we find ourselves in such a simple case? These are essentially technical questions for deontic logic, and I shall not pursue them here. In conclusion, it seems to me that the problem of possibly irresolvable moral conflict reveals serious flaws in the philosophical and semantic foundations of "orthodox" deontic logic, but also suggests a rich set of new problems and methods for such logic.

Notes

For support of this research the author is indebted to the John Simon Guggenheim Memorial Foundation.

In the course of writing this paper, I became acquainted with the following related work: C. L. Hamblin, "Quandaries and the Logic of Rules," *Journal of Philosophical Logic,* I (1972): 74–85; R. de Sousa, "The True and the Good," forthcoming; and T. Nagel, "War and Massacre," *Philosophy and Public Affairs* (1972, forthcoming). I wish also to acknowledge gratefully my debt to discussions and correspondence with A. al-Hibri, C. Daniels, R. de Sousa, H. Ishiguro, R. Stalnaker, and R. H. Thomason.

1. "The Logic of Conditional Obligation," to be published in the proceedings of a *Symposium on Exact Philosophy,* ed. M. Bunge; in the *Journal of Philosophical Logic,* I (1972): 417–38; and in the Synthese Library.

2. G. E. Moore, *Principia Ethica* (Cambridge: University Press, 1922), chapter 1, section 17, p. 25.

3. A. Sesonske, *Value and Obligation* (New York: Oxford, 1964), pp. 70 and 75.

4. See Count Jan Potocki, *The Saragossa Manuscript,* ed. R. Caillois, tr. E. Abbott (London: Cassell, 1962).

5. H. Ibsen, *A Doll's House,* Act 3.

6. F. H. Bradley, *Ethical Studies* (London: King, 1876), pp. 184–85.

7. J-P. Sartre, *L'Existentialisme est un Humanisme* (Paris: Nagel, 1964), p. 42.

8. Cf. his article on the Hungarian uprising, *Les Temps Modernes,* XII (November 1956): 579.

9. Comments on an earlier draft of this paper, presented at Cornell University, October 1971.

10. Bradley, *op. cit.,* chapter 4, p. 140.

8

Moral Dilemmas and Consistency in Ethics

Terrance C. McConnell

Recently it has been argued that there are genuine moral dilemmas
and that any theory which does not account for this fact is an
unrealistic one.[1] This represents a challenge to an assumption that
most moral theorists have held: an adequate ethical theory must not
allow for genuine moral quandaries. John Stuart Mill, for example,
in the last paragraph of the second chapter of *Utilitarianism,* seems
to be committed to such an assumption [pp. 54–55 in this volume].
Many others have also assented to this view.[2] The consensus among
those who hold this view seems to be that if a theory allows for
moral dilemmas then there is some sense in which it is incoherent
or inconsistent. Yet, oddly enough, the sense in which such a view
would be incoherent is rarely, if ever, spelled out. Put another way,
there seem to be no arguments for the belief that genuine moral
dilemmas must be ruled out. W. D. Ross does suggest that if the
same action were both morally required and forbidden, then "this
would be to put an end to all ethical judgment."[3] But how this

From the *Canadian Journal of Philosophy* 8 (1978): 269–87. Reprinted with the permission of the author and editors.

would put an end to all ethical judgment, Ross does not explain. Once one sees that few, if any, arguments have been advanced to support the commonly held assumption, one realizes that the recent challenges must be taken seriously.

Thus the main questions to which this paper is addressed are these: Must an adequate ethical theory allow for genuine moral dilemmas? Or must an adequate theory rule out such cases in order to avoid incoherence? I shall approach these questions by first spelling out two different senses in which our ethical reasoning might be thought to be inconsistent if there are genuine moral dilemmas. Discussing these two senses of inconsistency will cast light on the original questions. The conclusion that I shall eventually argue for is that we have good grounds for supposing that an adequate moral theory must rule out genuine dilemmas.

I

The first way in which the existence of moral dilemmas might be thought to lead to ethical inconsistency is this. The conjunction of three theses, each of which seems plausible and each of which is accepted by some philosophers, is inconsistent. The first thesis of the inconsistent triad is that there are genuine moral dilemmas. That is, there are situations in which an agent ought to do each of two things both of which he cannot do. If the situation is *genuinely* dilemmatic, then one is presented with two conflicting ought-claims and no further moral consideration is relevant to resolving the conflict. By contrast, a situation is merely apparently dilemmatic if two ought-claims conflict, but there are overriding moral reasons for acting on one rather than the other. Let us call the view that there are genuine moral dilemmas thesis (T1). The second thesis, (T2), is the view that 'ought' implies 'can.' This thesis states that the following is a principle of deontic logic: $OA \supset \Diamond A$. This principle is sometimes taken to be an axiom of deontic logic.[4] The third thesis is that the following principle of deontic logic holds: $(OA \& OB) \supset O(A \& B)$. This thesis, which I shall call (T3), is also an axiom of standard deontic logic.[5] The following argument, (A), shows that the conjunction of these three theses is inconsistent.[6]

(A) (i) OA premise
    (ii) OB premise

(iii) ~ ◇ (A & B)   premise
(iv) O(A & B) ⊃ ◇ (A & B)   premise
(v) (OA & OB) ⊃ O(A & B)   premise
(vi) O(A & B)   (i), (ii), (v), propositional calculus
(vii) ~ O(A & B)   (iii), (iv), propositional calculus

Lines (i), (ii), and (iii) represent thesis (T1). Theses (T2) and (T3) are set out in lines (iv) and (v), respectively. Or more accurately, lines (iv) and (v) are particular instances of these theses. Since lines (vi) and (vii) are contradictory, we know that the conjunction of the three is inconsistent. So at least one of the theses must be relinquished.

It is obvious that there are at least three solutions to this problem. Each of these solutions involves giving up one of the theses in question. Thus one may say that argument (A) forces us to give up the view that there are genuine moral dilemmas; that is, it shows that (T1) must be dropped. Or to put the point more cautiously, it shows that a condition of adequacy for any ethical theory is that it not allow for genuine moral dilemmas. One who defends this view will claim that lines (i)–(iii) cannot be jointly satisfied. The denial that there can be genuine moral dilemmas may be expressed as follows:

(ND)   (OA & OB) ⊃ ◇ (A & B)

Let us call this view *solution (1)*. Others will claim that (A) shows that 'ought' does not imply 'can.' If one omits line (iv) and line (vii), which depends on (iv), one will have a counterexample to this principle. This position I shall call *solution (2)*. *Solution (3)* is the view that the third thesis is the one that can be relinquished most reasonably. Defenders of this position will claim that argument (A), minus lines (v) and (vi), provides a counterexample to the deontic distribution principle. Adopting any one of the three solutions enables us to avoid the inconsistency, but it does so at the expense of forcing us to give up a thesis that at least some have found plausible. So if there are genuine moral dilemmas, then we are forced to give up one of two theses, each of which one has some reason to hold. In this sense one may regard the existence of apparent moral dilemmas as a challenge to the consistency of our current moral beliefs.

Argument (A) shows that standard deontic logic is indeed committed to ruling out the possibility of conflicting obligations. This same point can be made in another way, and in so doing it will be shown that (T1) leads to a second sense of ethical inconsistency. A principle, stated as an axiom in standard systems of deontic logic,[7] is the principle of deontic consistency (PC).

(PC) $OA \supset \, \sim O \sim A$

Intuitively (PC) just says that the same action cannot be obligatory and forbidden. To allow that the same act can be morally required and forbidden is not logically contradictory, but it does seem strange. If one accepts another principle of standard deontic logic,

(PD) $\square \, (A \supset B) \supset (OA \supset OB)$,

it can be shown that (T1) entails the denial of (PC).[8] So if there are genuine moral dilemmas, then our ethical reasoning is inconsistent in the sense that we are committed to both OA and $O \sim A$, propositions that are contraries according to (PC). And if we give up (PC), we shall also have to drop at least one of two other principles, each of which is accepted in most standard systems of deontic logic. The first is that if an action is obligatory then it is permissible. The second is that 'permissible' is definable by 'not ought not.'

(a) $OA \supset PA$
(b) $PA \equiv \, \sim O \sim A$

(a) and (b) entail (PC). So if moral dilemmas provide a counterexample to (PC), then at least one of these two principles must be given up too. Again we see that systems of standard deontic logic must rule out the possibility of genuine moral dilemmas. This shows that if (T1) is true, then our moral reasoning is radically different from what it is supposed to be by standard systems of deontic logic. I shall use the phrase 'the problem of moral dilemmas' as an abbreviated way of indicating that thesis (T1) represents a challenge to such basic moral principles as (T2), (T3), (PC), (a), and (b).

II

The problem set out in argument (A) will seem pressing only if each of the three theses is at least plausible. And one will need to worry about the adequacy of (PC) only if a good case can be made for (T1). I think that one can make a *prima facie* case for each of the three theses. I shall not document the reasons that have been given to support the principle that 'ought' implies 'can.' These are well

known, since this so-called Kantian principle is accepted by many.
It is, of course, not an uncontroversial principle. A number of different objections have been raised against it. Nonetheless the thesis
is held by many. One may wonder, though, why anyone would find
(T3) plausible. One can cite at least two reasons for supporting this
thesis. The first is simply that at the intuitive level (T3) seems plausible. If one ought to do each of two things, it seems quite natural
to think that one also ought to do both of them. Or if there are a
number of things each of which one ought to do, it is reasonable for
one to think that he ought to do all of these things. A second reason
for holding (T3) is this. Many have been struck by the close analogy
between the modal operator '□' and the deontic operator 'O', and
between the modal operator '◇' and the deontic operator 'P'. Since
obligation can be and has been construed as a kind of deontic or
moral necessity, the analogy is a natural one. Some even say that if
the analogue of the characteristic modal axiom, $\Box A \supset A$, is added to
deontic logic, the resulting systems are identical in form.[9] To illustrate briefly the parallels, the following analogous principles *do* hold
in standard systems of modal and deontic logic. (Here, of course,
the modal operators are to be understood as logical necessity and
possibility.)

(1) $(\Box A \vee \Box B) \supset \Box(A \vee B)$ (1') $(OA \vee OB) \supset O(A \vee B)$

(2) $\Box(A \supset B) \supset (\Box A \supset \Box B)$ (2') $O(A \supset B) \supset (OA \supset OB)$

(3) $\Diamond(A \vee B) \equiv (\Diamond A \vee \Diamond B)$ (3') $P(A \vee B) \equiv (PA \vee PB)$

(4) $\Diamond(A \& B) \supset (\Diamond A \& \Diamond B)$ (4') $P(A \& B) \supset (PA \& PA)$

In addition, the following analogous principles *do not hold* in standard systems of modal and deontic logic. (Since the counterexamples to these principles are well known. I shall not state them.)

(5) $\Box(A \vee B) \supset (\Box A \vee \Box B)$ (5') $O(A \vee B) \supset (OA \vee OB)$

(6) $(\Diamond A \& \Diamond B) \supset \Diamond(A \& B)$ (6') $(PA \& PB) \supset P(A \& B)$

The modal analogue of (T3), $(\Box A \& \Box B) \supset \Box(A \& B)$, clearly does
hold in standard systems of modal logic. So this gives one some
reason for believing that (T3) holds as well. And, as we have seen,
(T3) is an axiom of standard deontic logic. Though these reasons
are not conclusive ones, they surely show that the thesis is at least
plausible.

But is thesis (T1) plausible? At least three reasons have been
given for holding (T1). The first is that there are examples of situations that at least appear to be genuinely dilemmatic. One well

known example is discussed by Jean-Paul Sartre in his *Existentialism and Humanism*. This is the case of the student who believes that he ought to join the French forces and try to help defeat the German army. He also believes that he ought to stay at home and take care of his mother, who desperately needs his help. But clearly he cannot satisfy both of these ought-claims. Another example of an apparent moral dilemma occurs in Shakespeare's *Measure for Measure*. Angelo takes over the government of the city and immediately condemns to death one of his subjects, Claudio, for the crime of lechery. Isabella, Claudio's sister, goes to plead for her brother's life. She is a devout worshipper and a nun. Angelo tells her that he will free her brother only on the condition that she will sleep with him. As a sister and one devoted to her family, Isabella believes that she must do what is in her power to save her brother's life. As a nun, however, she is morally committed to preserving her virginity. Whatever she does, she believes that she will be doing something wrong.

A second reason for holding (T1) recognizes the importance of moral rules in our ordinary reasoning about ethical matters. It is always possible for two moral rules to conflict. And it is not just that it is logically possible that moral rules will conflict. If one looks at the complexity of the moral lives of most agents, one can hardly doubt that moral dilemmas will arise.[10] Most people take part in many different roles in society and are members of many different social groups. One incurs different obligations or duties as a friend, citizen, worker, spouse, etc. Given that each of us is involved in a complex network of relationships, it is very likely that some of the ought-claims binding on us will conflict and we will find ourselves in moral dilemmas on some occasions.

A third reason has been given to support the claim that there are genuine moral dilemmas. When agents face apparently dilemmatic situations they often experience regret after they act. Furthermore, in many cases this regret seems quite appropriate and certainly not irrational. Thesis (T1) provides a simple explanation for why a conscientious moral agent experiences this feeling and why it is not irrational. The agent experiences regret because he has failed to do something that he ought to have done. This provides evidence that the situation is genuinely dilemmatic because in many of these cases the agent sees that even if he had acted on the other of the conflicting alternatives he would still feel regret.[11] According to the advocates of solution (2) this regret shows that the agent ought to have done *both* of the incompatible things, though in fact he could not do both. Hence dilemmatic situations show that 'ought' does not imply 'can.' Defenders of solution (3) argue that this shows that the agent ought to have done *each* of the two things, though not both. Thus the existence of moral dilemmas forces us to reject (T3). But on either of these views, the regret that the agent experiences is taken as evidence for (T1).

It is fair to conclude that each of the three theses is at least plausible. Since this is the case, one can see why the problem of moral dilemmas is a worrisome one.

III

As I suggested earlier, I shall argue that the first solution is the most plausible one. If there are good reasons for saying that an adequate moral theory must not allow for genuine dilemmas, then the two kinds of ethical inconsistency described earlier will have been escaped. One will have avoided the problem to which argument (A) calls attention by showing that there are, after all, good reasons for giving up the first thesis of the inconsistent triad. The second problem will have been bypassed because one will not be forced to give up (PC) if (T1) is false.

Before proceeding, though, a preliminary remark needs to be made. In arguing that (T1) is false I am not necessarily ruling out the possibility that an agent can, by doing something forbidden, put himself in a situation where no matter what he does he will be doing something wrong. For example, one can make two promises that he knows conflict. Thus no matter what the agent does, he will break one of the promises. The situation arose, however, because the agent did something wrong. One might call situations like this dilemmatic; but if this were the only kind of dilemma that one could encounter, we would not be tempted to say that moral dilemmas show that our reasoning about ethical matters is incoherent. As one author suggests, "The existence of such cases is not morally disturbing, however, because we feel that the situation is not unavoidable: one had to do something wrong in the first place to get into it."[12] I shall, therefore, use the term 'genuine moral dilemma' to refer only to those quandaries, if there are any, which arise through no fault of the agent himself. Self-imposed moral predicaments may raise a number of philosophical problems (such as the problem of expressing contrary-to-duty imperatives in a standard deontic logic).[13] But if this were the only type of dilemma that was possible, one would not be likely to conclude that our moral reasoning is incoherent.

As we have seen, there are good reasons for holding (T1). The first task in defending solution (1) is to show that these reasons are not sufficient to guarantee the truth of (T1). Let us begin by considering the third reason given to support the view that there are genuine moral dilemmas. Is there any way to explain the appropriateness of the feeling of regret that an agent in a dilemmatic situation experiences without holding (T1)? To reiterate, the presence of regret is thought to show that the agent believes that he failed to do something that he ought to have done. And since he also believes the ought-claim that he acted on was binding on him, the situation was genuinely dilemmatic. It can be shown, however, that it does not follow that regret is appropriate only if the agent believes that he has failed to do something that he ought to have done. Regret, as it is ordinarily understood, is appropriate if some good has been lost, or if some bad, even if unavoidable, has obtained.[14] It is perfectly consistent and quite reasonable to say that an agent has done what he believes he ought, all things considered, to have done and feels regret. If one is in a situation where all available alternatives are in some sense bad and one did not create the situation by doing something wrong, then the case might be thought to be genuinely dilemmatic. If, however, there is a least evil alternative, it seems reasonable to say that there is one thing that the agent really ought to do, and the situation is not a genuine dilemma. In such circumstances doing the least evil act is surely the most rational thing to do, and one cannot regret having done the most rational thing. But one can regret being in a situation where only bad alternatives are open to one. That is, one can regret having to live in a world where such cases arise. The upshot of this is that the appropriateness of an agent's feeling regret after having acted in an apparently dilemmatic situation is not sufficient to show that the situation was genuinely dilemmatic.

The defender of (T1) may try to remedy this account by claiming that agents who act in apparently dilemmatic situations experience *remorse*, and not merely regret. Remorse, as a moral feeling, is appropriate only when the agent experiencing it believes that he has done something morally wrong.[15] If one assumes that remorse is the feeling that a conscientious agent should have after acting in an apparently dilemmatic situation, then (T1) will follow trivially.[16] It does so at considerable expense, however. This feeling (regret, remorse, or whatever) that a moral agent is expected to have is usually cited in a way that is supposed to provide an interesting *argument* for (T1). But if one assumes that the feeling the agent should have is remorse, one will have begged the question against the advocate of solution (1) (because the standard philosophical account of remorse says that the feeling is appropriate only if the agent has done something wrong). If, however, we *take* this feeling to be regret, then we can, as I have already shown, explain why agents in apparently dilemmatic situations have this feeling and why it is appropriate that they do. More importantly, that agents appropriately experience regret acknowledges that apparently dilemmatic situations are difficult. This difficulty is something that those who believe that there are genuine moral dilemmas want to emphasize. One can account for it, though, without holding (T1). So unless there are special reasons for taking this feeling to be remorse, it seems that we are justified in taking it to be regret. And if we are, the third reason is not sufficient to establish (T1).

We may now deal with the first two reasons for holding (T1). The first reason for believing that there are genuine moral dilemmas is that there are numerous examples of situations that certainly appear to be dilemmatic, like the case of Sartre's student or the predicament of Isabella in Shakespeare's play. And what will the defender of (T1) say to the person who believes that an adequate moral theory must not allow for genuine dilemmas? He will probably confront him with these examples and say, "If there are no genuine moral dilemmas, then what should the agent do in this situation?" But to suppose that the defender of solution (1) must always be able to answer this question is mistaken. It is not incumbent upon the advocate of solution (1) to supply the correct moral answer to every apparent quandary. The advocate of solution (1) may admit—and I shall develop this suggestion in more detail later—that there are cases in which we do not know what the agent ought, all things considered, to do. In other areas of inquiry, for example, history or physics, there may be some evidence supporting one hypothesis and some evidence supporting a conflicting hypothesis. That one does not know which hypothesis is correct does not *by itself* cast doubt on the claim that there is a uniquely correct answer to the question at issue. So too the admission that one does not know what an agent in a conflict situation really ought to do does not by itself cast doubt on solution (1). Since this admission can be made, the first reason given to support (T1) is not sufficient.

The second reason cited to show that (T1) is true is that the sources of one's obligations and duties are numerous and diverse, and it is highly probable that these will come into conflict. The advocate of solution (1) cannot, of course, deny that obligations can and do sometimes conflict. That is, there are sometimes cases where each of two ought-claims *seems* to apply to a situation but both cannot be satisfied. It does not follow from this, however, that there are genuine moral dilemmas. No one will deny that there are cases of moral conflict where one ought-claim clearly overrides the other. One such case is when one's obligation to help an accident victim overrides one's obligation to meet a friend for lunch (as one promised to do). One may regard this as a paradigm case of one ought-claim overriding another.[17] Since there are cases where the overriding relationship does hold, the *mere* fact that two ought-claims can or do conflict does not show that there are genuine moral dilemmas. And this is so even if one does not know in some particular case which ought-claim takes precedence. The overriding relationship may still hold such cases.

I have, up to this point, argued for a fairly weak claim: viz., the reasons usually given to support (T1) do not guarantee the truth of that thesis. A defender of (T1), however, might well grant this. He might claim, though, that the burden of proof is still on the advocate of solution (1). The reason for this, he might argue, is that he has shown that there *appear* to be genuine moral dilemmas. The first two reasons that he gives to support (T1) show this. True, he will admit, it is possible that in every conflict case one ought-claim overrides the other. But since in many cases we do not know which, if either, overrides the other, appearances support the view that there are genuine moral quandaries. In short, the burden of proof will be on one who claims that these appearances are deceptive. This response is, I think, correct. It shows that the defender of solution (1) has more work to do; he needs to give some reason(s) for believing that (T1) is false.

IV

There are two phenomena that are frequently associated with dilemmatic situations. One is that agents facing apparent quandaries frequently *seek moral advice*. Sartre's student is a case in point. It is important to note that the student sought advice because each of two moral claims seemed equally incumbent on him, and he could not jointly satisfy them. The second phenomenon is that after acting on one or the other of the alternatives, agents in these apparently dilemmatic situations often *experience moral doubt.* They wonder if they have done the right thing; or more typically, they worry that they have done the wrong thing. And not only do agents in these situations seek moral advice and experience moral doubt, but in addition in many of these cases we are ordinarily prepared to say that such behaviour is appropriate, reasonable, and maybe even expected. That such behaviour is regarded as appropriate can easily be explained by the advocate of solution (1). By contrast, it seems that advocates of (T1) cannot adequately account for this fact. Let me explain why this is the case.

Suppose that an agent who is in a situation such that it seems that he ought to do each of two things both of which he cannot do asks another person for moral advice. If the person whom he asks believes that there are genuine moral dilemmas and that the agent is in one, then this person will simply advise the agent that he ought to do each (or both)[18] of the two actions. The agent would hardly consider this a satisfactory answer. He would not have asked for advice had he anticipated this answer. His asking for advice indicates that he believes that there is some one thing that he ought to do and he is trying to find out what that is. And since we ordinarily regard such behaviour as appropriate, it seems that we too believe that there is some one thing that the agent ought to do. The behaviour is appropriate because the agent does not know what he really ought to do. But if the advocate of (T1) is correct, the agent does know what he ought, everything considered, to do. He ought to do each (or both) of the actions. It seems, then, that the defenders of (T1) cannot claim that the agent is genuinely trying to discover what is really morally required of him. Since (according to them) the agent already knows what he ought to do, they are committed to saying that such behaviour is irrational. Either that or they will have to explain the behaviour as being something other than what it appears to be.

Moral doubt is also often associated with apparently dilemmatic situations. After acting in such situations a person will often ask, "Was what I did right?" Depending on the seriousness of the apparent dilemma, the person may worry about this for a long time, especially if certain undesirable consequences of his actions are frequently evident to him. One can imagine the doubt that Isabella might have experienced had Claudio actually been executed. And if

Sartre's student had decided to join the French forces and his mother had subsequently died, the doubt he would have experienced would probably have been long-lasting. Furthermore, in cases like these we think that this doubt is appropriate and certainly not irrational. If one holds (T1) (and also believes that the agent in question has just faced a genuine dilemma), then the answer to these questions that the agent raises will be obvious. He will say that of course the agent failed to do at least one thing that he ought to have done. He may say that the agent is not blameworthy, assuming that he did not get himself into the predicament by doing something forbidden. But as long as the agent recognized the force of each of the ought-claims *before* he acted, then it seems that the advocate of (T1) cannot take these questions seriously. He will have to explain them as being something other than what they appear to be. He may, for example, say that the agent is involved in some sort of self-deception or act of bad faith, and is merely trying to get someone to persuade him that he did nothing wrong. Whatever he says, however, it seems that he cannot allow that this experience is genuine moral doubt. Yet I think that most of us want to say that in many such cases this moral doubt is genuine. If it is, then it seems that the advocate of (T1) will be hard pressed to explain it. That he seems to be forced to treat this phenomenon as something other than what it appears to most of us to be is a weakness of his view.

What I have shown so far is that there are two phenomena commonly associated with dilemmatic situations that seem either inexplicable, irrational, or something other than what they appear to be if one holds that (T1) is true. But suppose that one denies (T1). Can the person who opts for solution (1) explain these two phenomena adequately? I think that he can. There is a way of looking at and explaining apparent moral dilemmas which is both plausible and open to the advocate of solution (1). An apparently dilemmatic situation is one where two (or more) ought-claims seem to be applicable to one's situation, one cannot act on both of these, and one *does not know* on which of these one ought, everything considered, to act. If this is a plausible account of apparent moral dilemmas, then one can support solution (1) and still explain the phenomena that it seems the advocate of (T1) cannot adequately explain. The first phenomenon, that moral agents frequently ask what they ought to do even after they learn that two moral rules or principles seem to be applicable to their situation, is easily explained. An agent asks the question because he does not know what he really ought to do. The case is one of *moral conflict,* but *not* a genuine moral dilemma (thus I use the term 'moral conflict' in a somewhat technical sense), simply because each of two incompatible ought-claims seems legitimate and one does not know on which one ought, all things considered, to act. The agent believes that there is one thing that he really ought to do, but he does not know what it is. That after acting in apparently dilemmatic situations agents often experience moral doubt also can be explained by the advocate of solution (1). When a person is in such a situation, he frequently must act within a limited period of time. So he may well be forced to choose between the conflicting alternatives even though he does not know on which he ought to act. If at the time of his action he was not sure what he really ought to do, it is quite reasonable that after the fact he wonders if what he did was what he really should have done. Of course, doubt may arise even if the agent was not uncertain before acting. In such cases the agent does not recognize the conflict-nature of his situation until after he has acted.

In section III I showed that giving up thesis (T1) is not as costly as one might think. One can, contrary to what some of the advocates of solutions (2) and (3) think, account for why it is not irrational for agents who have faced apparently dilemmatic situations to experience regret even if (T1) is false. In this section I have argued that unless we support solution (1) we cannot account for the appropriateness of an agent's seeking moral advice before acting (in an apparently dilemmatic situation) or for experiencing moral doubt after acting.

V

If my argument is correct, then at the very least the burden of proof has been shifted to those who claim that there are genuine moral dilemmas. Put another way, a *prima facie* case for solution (1) has been made. Unfortunately, though, there is what appears to be a fatal objection to this argument. The person who claims that there are genuine moral dilemmas does, it seems, have a way of accounting for the appropriateness of an agent's seeking moral advice and experiencing moral doubt. The person who holds (T1) can draw a distinction between cases of *actual* moral dilemmas and cases that are *only apparently* dilemmatic. He can allow, that is, that in some cases when two ought-claims conflict there are moral grounds for saying that one takes precedence over the other. And he may even grant that such grounds exist when at first glance this does not appear to be the case. What he cannot allow, of course, is that in every conflict case there is a morally preferable alternative. The importance of this distinction is this. The defender of (T1) can admit that when one is in an apparent moral dilemma he cannot be sure that the situation is genuinely dilemmatic. Since this is the case, it is quite reasonable for (and perhaps even required of) the agent to seek moral advice. If the situation is not genuinely dilemmatic, others may be able to help the agent discover what he really ought to do. Even if the situation turns out to be genuinely dilemmatic, it may be that one ought to take all reasonable precautions before acting. So even if one holds that there are genuine moral dilemmas, it seems that he can account for the appropriateness of an agent's seeking moral advice in such circumstances.

The same line of reasoning applies to experiencing moral doubt after the fact. If one has to act in a situation that is apparently dilemmatic but may not actually be so, one might needlessly do something wrong. If the situation is a genuine dilemma, then one cannot help but do at least one wrong thing. If, however, the situation is merely an apparent dilemma, then there is just one thing that the agent really ought to do. Moral doubt is appropriate because the agent *might* have done something wrong when he could have avoided doing so.[19] So the defender of (T1) can give essentially the same explanation of an agent's seeking moral advice and experiencing moral doubt as the advocate of solution (1) can give.

It seems, then, that the person who says that there are genuine moral dilemmas can get around what initially appeared to be a very forceful argument against (T1). This response, however, is not without some difficulties. The key to the move made by the defender of (T1) is his being able to distinguish moral dilemmas from those that are only apparently dilemmatic. An important question must be asked, though. Does the advocate of (T1) have a criterion (in the epistemic sense) for distinguishing situations that are truly dilemmatic from those that erroneously appear so? Notice that if he did have a plausible criterion he would, in effect, have a straightforward argument for (T1). I know of no such criterion, however. In fact, it seems reasonable to assume that there is no such criterion; even the defender of (T1) can and must grant this. (He must grant it in order to explain why advice-seeking and the experiencing of doubt are appropriate.)

If there is no criterion for picking out genuine dilemmas from apparent ones, then one must raise another question. Given that an agent is facing a situation that at least appears to be a dilemma, when will it be rational for him to seek advice and experience doubt? There seem to be at least two different responses that the defender of (T1) might make. First, he might claim that advice-seeking and experiencing doubt are *always* appropriate because it is impossible for one to tell whether he is in an apparent dilemma or a genuine one. To take this line is to admit that one must treat every case *as if* it were only an apparent dilemma. But this puts the defender of (T1) in a very weak position. If he admits this much, one may wonder what grounds there are for ever thinking that one is in a genuine dilemma. It seems that there are none. To grant that we must presuppose that each case we face is only an apparent dilemma is surely to give the advocate of solution (1) all that he needs. After all, the advocate of (T1) is recommending that we behave as if solution (1) were correct. Admittedly this is compatible with (T1). But the strongest claim that a defender of this thesis could make is the following: looking back, as it were, over a whole range of cases, one may say that some of these cases were genuinely dilemmatic; however, one cannot correctly identify the real dilemmas. But even this innocuous version of the first thesis will not be defensible. One will not be able to give any reasons to support (T1). Notice, for example, that one will no longer be able to point to examples of apparent dilemmas as a reason for holding (T1). Any case to which one points will be one where seeking moral advice and experiencing moral doubt are appropriate because one *does not* know (according to the advocate of (T1) himself) what he really ought to do. By his own admission, then, there are no situations that are known to be dilemmatic. That there are genuine dilemmas must, it seems, be accepted on faith. Surely this first response is not the one that the defender of (T1) will want to make.

Suppose, however, that the defender of (T1) takes a second line. Suppose he says that seeking advice and experiencing doubt in apparently dilemmatic situations are *not always* rational. He may say, for example, that if an agent in an apparently dilemmatic situation has good reasons to believe that it is a genuine dilemma (though he can never know for sure), then seeking moral advice and experiencing doubt will be irrational. In these cases, he might say, resignation or a Stoic-like attitude is more appropriate. One immediate problem with this response is spelling out what will count as *good reasons* for believing that one's situation is genuinely dilemmatic, and it is not easy to imagine what the defender of (T1) might say. Perhaps one should merely "take a second look" at the available data. If the situation still appears to be dilemmatic, then one may assume that it is really dilemmatic. It may be that one should take all reasonable precautions, but doing so need not include going out of one's way to seek moral advice or to experience doubt. The problem with taking this line is that it will render as inappropriate in far too many cases behaviour that most of us would not regard as such. Surely the ongoing doubt that Sartre's student is likely to experience is not irrational. And Isabella's seeking moral advice after much pondering (taking several "second looks") does not seem inappropriate. Of course, most of us *may* be wrong in thinking that this doubt is not irrational or that this advice-seeking is not inappropriate. But since the advocate of (T1) must treat these cases in a way that does not seem proper, it is up to him to show that these appearances are deceptive.

The first thesis, then, conflicts (in a fundamental way) with the way that we regard situations of moral conflict. If one defending this thesis says that seeking moral advice and experiencing moral doubt are always appropriate when one is facing an apparent predicament, then he will be giving up too much to the defender of solution (1). If he says that such behaviour is not always appropriate, then he must spell out under what conditions it is inappropriate. And as we have seen, the most natural way of spelling this out commits him to treating some behaviour as inappropriate when we would ordinarily regard it as quite reasonable.[20] We now can see that the objection to the argument put forth in section IV is not one that the defender of (T1) can press. Given this, there are good grounds for believing that (T1) is false.

VI

In spite of the above argument, some may still want to claim that there are genuine moral dilemmas. Many who hold (T1) do so because they believe that those who advocate solution (1) must present an unrealistic picture of the moral life. In particular, to deny that there are moral dilemmas is to be much too optimistic about the moral life. One author, referring to a version of what I call solution (1), says, "This view seems to me to make the moral life too easy."[21] Though this is a common reaction to the thesis that an adequate ethical theory cannot allow for genuine dilemmas, it is nonetheless a mistaken one. The advocate of solution (1) is not committed to a view that depicts the moral life as too easy. In fact, I have already shown that one who supports solution (1) can explain why the moral life is difficult. When an agent is in an apparently dilemmatic situation he has some reason to believe that he ought to do each of two things both of which he cannot do. He does not know, however, which of these things he ought, all things considered, to do. He will necessarily fail to do at least one of the things, and some sort of harm will ensue. (This is why regret is appropriate in such situations.) Many apparently dilemmatic situations leave one little time to contemplate or deliberate. The difficulty, then, is that an agent may have to act without knowing what he really ought to do.[22] Even if the agent had more time to deliberate, the situation may be so difficult that he still will not know what to do. It is because of this feature of the moral life that an agent's seeking moral advice or experiencing moral doubt is appropriate. And unless we are prepared to say that such behaviour is inappropriate, we are committed to believing that (T1) is false. So, ironically, some of the difficult aspects of the moral life can be accounted for only if we assume that there are no genuine moral dilemmas.

Some critics have asserted that it is no comfort to an agent facing an apparently dilemmatic situation to know that there is, in principle, a morally correct solution to his problem even though it is currently unknown to us.[23] The agent must still face a difficult moral problem. To cite this as an objection to solution (1), however, is to misunderstand the purpose of that view. Such a critic seems to be attributing to the advocate of solution (1) the view that dilemmatic situations only appear to be difficult, and if the agent just reflects he will see that the difficulty is only apparent. Solution (1), however, does not entail this. The difficulty involved in apparently dilemmatic situations is *not illusory,* as we have just seen. Solution (1) was never intended to comfort moral agents facing conflict situations. If it is intended to comfort anyone it is the moral theorist who is worried—or at least should be once he sees the problem of moral dilemmas—about the consistency of our moral reasoning. Not only is solution (1) not intended to comfort moral agents facing apparent quandaries, but it explicitly calls attention to the difficulty of such cases. That solution (1) gives no comfort to agents in conflict situations is not a weakness of the view; it is rather an indication of the realistic picture of the moral life that it presents.

What I have tried to do in this paper is to defend an assumption which has been held, but not really argued for, by many ethical theorists: an adequate moral theory must not allow for genuine dilemmas. The most important result of this argument is that it provides a solution to the problem set out in argument (A). We do have good reasons for giving up the first thesis of the inconsistent triad. And this allows us to retain the most basic principles of deontic logic as well, viz., (PC), (a), and (b).[24] We may conclude, then, at least with respect to the problem of moral dilemmas, that our basic ethical reasoning is not incoherent.

Notes

I am indebted to a number of people for many helpful suggestions. Among those to whom a special thanks is owed are Norman Dahl, Barry Hoffmaster, Gary Iseminger, Husain Sarkar, Chris Swoyer, and a referee of the *Canadian Journal of Philosophy*.

1. Those who have argued for or asserted this position include the following: E. J. Lemmon, "Moral Dilemmas," *Philosophical Review* 71 (1962), pp. 139–58 [chapter 5 in this volume], Bernard Williams, "Ethical Consistency," *Proceedings of the Aristotelian Society,* Supplementary Volume 39 (1965), pp. 103–24 [chapter 6 in this volume], and *Morality: An Introduction to Ethics* (Harper & Row, 1972); Roger Trigg, "Moral Conflict," *Mind* 80 (1971), pp. 41–55; Bas C. van Fraassen, "Values and the Heart's Command." *Journal of Philosophy* 70 (1973), pp. 5–19 [chapter 7 in this volume]; and P. H. Nowell-Smith, "Some Reflections on Utilitarianism," *Canadian Journal of Philosophy* 2 (1972–73), pp. 417–31.

2. To mention just a few who hold this view, see David Lyons, *Forms and Limits of Utilitarianism* (Oxford University Press, 1965), p. 21; John Rawls, *A Theory of Justice* (Harvard University Press, 1971), pp. 133–34; and Hector-Neri Castañeda, *The Structure of Morality* (Charles C. Thomas, 1974).

3. W. D. Ross, *Foundations of Ethics* (Oxford University Press, 1939), p. 60. Ross also claims (p. 86) that *simply* drawing the distinction between *prima facie* obligations and actual ones shows that the problem of moral dilemmas is nonexistent. It will be clear, however, that merely draw-

ing this distinction does not solve the problem of moral dilemmas as I shall set it out; it rather presupposes what the solution is.

4. See Brian F. Chellas, "Conditional Obligation," in S. Stenlund, ed., *Logical Theory and Semantic Analysis* (D. Reidel Publishing Co., 1974), p. 23.

5. See Dagfinn Føllesdal and Risto Hilpinen, "Deontic Logic: An Introduction," in Hilpinen, ed., *Deontic Logic: Introductory and Systematic Readings* (D. Reidel Publishing Co., 1971), p. 13.

6. An informal account of this argument is presented by Bernard Williams in "Ethical Consistency," p. 118 [p. 130 in this volume]. Throughout this work 'OA' is to be read 'X (the agent to whom the ought-claim is addressed) ought, all things considered, to do A'. The qualification 'all things considered' indicates that the ought-claim is not merely a *prima facie* one. 'PA' is to be read 'X is permitted to do A'. The logical connectives are to be understood in the usual way. However, the modal operators should not be taken to stand for logical possibility and necessity (unless otherwise noted). The 'can' in the principle that 'ought' implies 'can' usually involves a notion stronger than mere logical possibility; the same is true of the 'cannot' in the assertion that there are moral dilemmas. As a result the modal operators should be taken to stand for something like *physical* possibility and necessity.

7. See Føllesdal and Hilpinen, "Deontic Logic: An Introduction," p. 13.

8. Even those who argue that there are moral dilemmas accept (PD). See, for example, E. J. Lemmon, "Deontic Logic and the Logic of Imperatives," *Logique et Analyse* 8 (1965), p. 40, and van Fraassen, "Values and the Heart's Command," p. 15 [p. 149 in this volume]. The argument which shows that the advocate of (T1) is committed to giving up (PC) is a simple one and is set out in my "Moral Dilemmas and Requiring the Impossible," *Philosophical Studies* 29 (1976), pp. 410–11.

9. See William H. Hanson, "Semantics for Deontic Logic," *Logique et Analyse* 8 (1965), p. 178.

10. This point is made by R. W. Beardsmore in his *Moral Reasoning* (Routledge & Kegan Paul, 1969), p. 111.

11. Several authors have suggested that this argument shows that there are genuine moral dilemmas. See, for example, Bernard Williams, "Ethical Consistency," pp. 109–13 [pp. 121–25 in this volume], and Roger Trigg, "Moral Conflict," pp. 47–52. Williams uses this argument to try to establish what I call solution (3); Trigg, to establish solution (2).

12. Thomas Nagel, "War and Massacre," *Philosophy & Public Affairs* 1 (1972), p. 143.

13. This problem is set out by Roderick Chisholm in his "Contrary-to-duty Imperatives and Deontic Logic," *Analysis* 24 (1963), pp. 33–36.

14. John Rawls, *A Theory of Justice,* p. 422 and pp. 442–44.

15. See P. H. Nowell-Smith, *Ethics* (Penguin Books, 1954), p. 26.

16. This needs to be qualified. It must also be the case that the agent still regards the alternative that he acted on as morally required. An agent might, in retrospect, believe that there was a morally preferable alternative in the situation that he faced. If the agent acted on the alternative that he now believes was morally wrong, then remorse would be appropriate even though the situation was not genuinely dilemmatic. Remorse might also be appropriate if the agent put himself in the dilemmatic situation by doing something forbidden, such as knowingly making conflicting promises.

17. It would be helpful to have an account of what is involved when one ought-claim overrides or has more moral weight than another ought-claim. No such account is presented here. One very provocative attempt to elucidate this concept is presented by Robert Nozick, in his "Moral Complications and Moral Structures," *Natural Law Forum* 13 (1968) pp. 1–50.

18. The advocate of solution (2) will advise him that he ought to do *both;* the defender of solution (3), that he ought to do *each.*

19. If in moral matters we held a view something like "let bygones be bygones," then moral doubt would not be appropriate here. But we do not hold such a view. We have such notions as duties of reparation or duties to make amends, and if a person did the wrong thing in a situation that was only apparently dilemmatic he may well incur some such duty.

20. There may, of course, be a way of spelling out the second line of reasoning which does not commit the advocate of (T1) to these undesirable consequences. But until a different alternative is suggested, the criticism stands. I do not see any obvious way to alter this second line of reasoning to make the view more plausible.

21. E. J. Lemmon, "Deontic Logic and the Logic of Imperatives," p. 45.

22. Some may think that because an agent must sometimes act without knowing what he really ought to do, this shows that 'ought' does not imply 'can.' But in such situations the agent surely *can* do what he ought to do in many senses of the term, including both the physical and logical senses.

23. This objection is presented by Bas C. van Fraassen, "Values and the Heart's Command, " page 12 [p. 145 in this volume], and Michael Walzer, "Political Action: The Problem of Dirty Hands," *Philosophy & Public Affairs* 2 (1972–73), p. 161.

24. One may wonder just how basic (PC) is. E.J. Lemmon, in "Deontic Logic and the Logic of Imperatives," p. 51, claims that if (PC) does not hold in a system of deontic logic, then all that remains are truisms and paradoxes. If this claim is correct, then there are obvious theoretical advantages in retaining (PC).

9

The Fragmentation of Value

Thomas Nagel

I want to discuss some problems created by a disparity between the fragmentation of value and the singleness of decision. These problems emerge in the form of practical conflicts, and they usually have moral components.

By a practical conflict I do not mean merely a difficult decision. Decisions may be difficult for a number of reasons: because the considerations on different sides are very evenly balanced; because the facts are uncertain; because the probability of different outcomes of the possible courses of action is unknown. A difficult choice between chemotherapy and surgery, when it is uncertain which will be more effective, is not an example of what I mean by practical conflict, because it does not involve conflict between values which are incomparable for reasons apart from uncertainty about the facts. There can be cases where, even if one is fairly sure about the outcomes of alternative courses of action, or about their probability distributions, and even though one knows how to distinguish the pros and cons, one is nevertheless unable to bring them together in

From *Mortal Questions* (Cambridge: Cambridge University Press, 1979), pp. 128–41. Reprinted with the permission of the publisher.

a single evaluative judgment, even to the extent of finding them evenly balanced. An even balance requires comparable quantities.

The strongest cases of conflict are genuine dilemmas, where there is decisive support for two or more incompatible courses of action or inaction. In that case a decision will still be necessary, but it will seem necessarily arbitrary. When two choices are very evenly balanced, it does not matter which choice one makes, and arbitrariness is no problem. But when each seems right for reasons that appear decisive and sufficient, arbitrariness means the lack of reasons where reasons are needed, since either choice will mean acting against some reasons without being able to claim that they are *outweighed.*

There are five fundamental types of value that give rise to basic conflict. Conflicts can arise within as well as between them, but the latter are especially difficult. (I have not included self-interest in the group; it can conflict with any of the others.)

First, there are specific obligations to other people or institutions; obligations to patients, to one's family, to the hospital or university at which one works, to one's community or one's country. Such obligations have to be incurred, either by a deliberate undertaking or by some special relation to the person or institution in question. Their existence depends in either case on the subject's relation to others, although the relation does not have to be voluntary. (Even though young children are not at liberty to choose their parents or guardians, parental care creates some obligation of reciprocal future concern.)

The next category is that of constraints on action deriving from general rights that everyone has, either to do certain things or not to be treated in certain ways. Rights to liberty of certain kinds, or to freedom from assault or coercion, do not depend on specific obligations that others have incurred not to interfere, assault, or coerce. Rather, they are completely general, and restrict what others may do to their possessor, whoever those others may be. Thus a doctor has both specific obligations to his patients and general duties to treat anyone in certain ways.

The third category is that which is technically called utility. This is the consideration that takes into account the effects of what one does on everyone's welfare—whether or not the components of that welfare are connected to special obligations or general rights. Utility includes all aspects of benefit and harm to all people (or all sentient beings), not just those to whom the agent has a special relation, or has undertaken a special commitment. The general benefits of medical research and education obviously come under this heading.

The fourth category is that of perfectionist ends or values. By this I mean the intrinsic value of certain achievements or creations, apart from their value *to* individuals who experience or use them. Examples are provided by the intrinsic value of scientific discovery, of artistic creation, of space exploration, perhaps. These pursuits do of course serve the interests of the individuals directly involved in them, and of certain spectators. But typically the pursuit of such ends is not justified solely in terms of those interests. They are thought to have an intrinsic value, so that it is important to achieve fundamental advances, for example, in mathematics or astronomy even if very few people come to understand them and they have no practical effects. The mere existence of such understanding, somewhere in the species, is regarded by many as worth substantial sacrifices. Naturally opinions differ as to what has this kind of worth. Not everyone will agree that reaching the moon or Mars has the intrinsic value necessary to justify its current cost, or that the performance of obscure or difficult orchestral works has any value apart from its worth to individuals who enjoy them. But many things people do cannot be justified or understood without taking into account such perfectionist values.

The final category is that of commitment to one's own projects or undertakings, which is a value in addition to whatever reasons may have led to them in the first place. If you have set out to climb Everest, or translate Aristotle's *Metaphysics,* or master the *Well-Tempered Clavier,* or synthesize an amino acid, then the further pursuit of that project, once begun, acquires remarkable importance.[1] It is partly a matter of justifying earlier investment of time and energy, and not allowing it to have been in vain. It is partly a desire to be the sort of person who finishes what he begins. But whatever the reason, our projects make autonomous claims on us, once undertaken, which they need not have made in advance. Someone who has determined to master the *Well-Tempered Clavier* may say 'I can't go to the movies, I have to practice'; but it would be strange for him to say that he had to master the *Well-Tempered Clavier.*

These commitments should not be confused with self-interest, for self-interest aims at the integrated fulfillment over time of *all* one's interests and desires (or at least those desires one does not wish to eliminate). Special commitments may, in their pursuit, be inimical to self-interest thus defined. They need not have been undertaken for self-interested reasons, and their pursuit certainly need not be controlled by self-interest.

Obligations, rights, utility, perfectionist ends, and private commitments—these values enter into our decisions constantly, and conflicts among them, and within them, arise in medical research, in politics, in personal life, or wherever the grounds of actions are not artificially restricted. What would it mean to give a system of priorities among them? A simpler moral conception might permit a solution in terms of a short list of clear prohibitions and injunctions, with the balance of decision left to personal preference or discretion, but that will not work with so mixed a collection. One might try to order them. For example: never infringe general rights, and undertake only those special obligations that cannot lead to the infringement of anyone's rights; maximize utility within the range of action left free by the constraints of rights and obligations; where utility would be equally served by various policies, determine the choice by reference to perfectionist ends; and finally, where this leaves anything unsettled, decide on grounds of personal commitment or even simple preference. Such a method of decision is absurd, not because of the particular order chosen but because of its absoluteness. The ordering I have given is not arbitrary, for it reflects a degree of relative stringency in these types of values. But it is absurd to hold that obligations can never outweigh rights, or that utility, however large, can never outweigh obligation.

However, if we take the idea of outweighing seriously, and try to think of an alternative to ordering as a method of rationalizing decision in conditions of conflict, the thing to look for seems to be a single scale on which all these apparently disparate considerations can be measured, added, and balanced. Utilitarianism is the best example of such a theory, and interesting attempts have been made to explain the apparent priority of rights and obligations over utility in utilitarian terms. The same might be tried for perfectionist goals and personal commitments. My reason for thinking that such explanations are unsuccessful, or at best partially successful, is not just that they imply specific moral conclusions that I find intuitively unacceptable (for it is always conceivable that a new refinement of the theory may iron out many of those wrinkles). Rather, my reason for doubt is theoretical: I do not believe that the source of value is unitary—displaying apparent multiplicity only in its application to the world. I believe that value has fundamentally different kinds of sources, and that they are reflected in the classification of values into types. Not all values represent the pursuit of some single good in a variety of settings.

Think for example of the contrast between perfectionist and utilitarian values. They are *formally* different, for the latter takes into account the number of people whose interests are affected, and the former does not. Perfectionist values have to do with the mere level of achievement and not with the spread either of achievement or of gratification. There is also a formal contrast between rights or obligations and any ends, whether utilitarian or perfectionist, that are defined in terms of the outcome of actions—in terms of how things are as a result. The claims represented by individual obligations begin with relations betwen individuals, and although the maintenance of those relations in a satisfactory form must be part of any utilitarian conception of a good state of affairs, that is not the basic motive behind claims of obligation. It may be a good thing that people keep their proimses or look after their children, but the reason a person has to keep his own promises is very different from the reason he has to want other people unconnected with him to keep their promises—just because it would be a good thing, impersonally considered. A person does not feel bound to keep his promises or look after his children because it would be a good thing, impersonally considered. There certainly are things we do for such reasons, but in the motive behind obligations a more personal outlook is essential. It is your *own* relation to the other person or the institution or community that moves you, not a detached concern for what would be best overall.

Reasons of this kind may be described as agent-centered or subjective (though the term 'subjective' here should not be misunderstood—it does not mean that the general principles of obligation are matters of subjective preference which may vary from person to person). The reasons in each case apply primarily to the individual involved, as reasons for *him* to want to fulfil his obligations—even though it is also a good thing, impersonally considered, for him to do so.

General rights are less personal in their claims, since a right to be free from interference or assault, for example, does not derive from the possessor's relation to anyone in particular: everyone is obliged to respect it. Nevertheless, they are agent-centered in the sense that the reasons for action they provide apply primarily to individuals whose actions are in danger of infringing such rights. Rights mainly provide people with reasons not to do certain things to other people—not to treat them or interfere with them in certain ways. Again, it is objectively a good thing that people's rights not be violated, and this provides disinterested parties with some reason for seeing that X's rights are not violated by Y. But this is a secondary motive, not so powerful as the reason one has not to violate anyone's rights directly. (That is why it is reasonable for defenders of civil liberties to object to police and judicial practices that violate the rights of criminal suspects, even when the aim of those policies is to prevent greater violations by criminals of the rights of their victims.) In that sense the claims deriving from general rights are agent-centered: less so than those deriving from special obligations, but still definitely agent-centered in a sense in which the claims of utility or perfectionist ends are not. Those latter claims are impersonal or outcome-centered; they have to do with what happens, not, in the first instance, with what one does. It is the contribution of what one does to what happens or what is achieved that matters.

This great division between personal and impersonal, or between agent-centered and outcome-centered, or subjective and objective reasons, is so basic that it renders implausible any reductive unification of ethics—let alone of practical reasoning in general. The formal differences among these types of reasons correspond to deep differences in their sources. We appreciate the force of impersonal reasons when we detach from our personal situation and our special relations to others. Utilitarian considerations arise in this way when our detachment takes the form of adopting a general point of view that comprehends everyone's view of the world within it. Naturally the results will not always be clear. But such an outlook is obviously very different from that which appears in a person's concern for his special obligations to his family, friends, or colleagues. There he is thinking very much of his particular situation in the world. The two motives come from two different points of view, both important, but fundamentally irreducible to a common basis.

I have said nothing about the still more agent-centered motive of commitment to one's own projects, but since that involves one's own life and not necessarily any relations with others, the same points obviously apply. It is a source of reasons that cannot be assimilated either to utility, or perfectionism, or rights, or obligations (except that they might be described as obligations to oneself).

My general point is that the formal differences among types of reason reflect differences of a fundamental nature in their sources, and that this rules out a certain kind of solution to conflicts among these types. Human beings are subject to moral and other motivational claims of very different kinds. This is because they are complex creatures who can view the world from many perspectives—individual, relational, impersonal, ideal, etc.—and each perspective presents a different set of claims. Conflict can exist within one of these sets, and it may be hard to resolve. But when conflict occurs between them, the problem is still more difficult. Conflicts between personal and impersonal claims are ubiquitous. They cannot, in my view, be resolved by subsuming either of the points of view under the other, or both under a third. Nor can we simply abandon any of them. There is no reason why we should. The capacity to view the world simultaneously from the point of view of one's relations to others, from the point of view of one's life extended through time, from the point of view of everyone at once, and finally from the detached viewpoint often described as the view *sub specie aeternitatis* is one of the marks of humanity. This complex capacity is an obstacle to simplification.

Does this mean, then, that basic practical conflicts have no solution? The unavailability of a single, reductive method or a clear set of priorities for settling them does not remove the necessity for making decisions in such cases. When faced with conflicting and incommensurable claims we still have to do something—even if it is only to do nothing. And the fact that action must be unitary seems to imply that unless justification is also unitary, nothing can be either right or wrong and all decisions under conflict are arbitrary.

I believe this is wrong, but the alternative is hard to explain. Briefly, I contend that there can be good judgment without total justification, either explicit or implicit. The fact that one cannot say why a certain decision is the correct one, given a particular balance of conflicting reasons, does not mean that the claim to correctness is meaningless. Provided one has taken the process of practical justification as far as it will go in the course of arriving at the conflict, one may be able to proceed without further justification, but without irrationality either. What makes this possible is *judgment*—essentially the faculty Aristotle described as practical wisdom, which reveals itself over time in individual decisions rather than in the enunciation of general principles. It will not always yield a solution: there are true practical dilemmas that have no solution, and there are also conflicts so complex that judgment cannot operate confidently. But in many cases it can be relied on to take up the slack that remains beyond the limits of explicit rational argument.

This view has sometimes been regarded as defeatist and empty since it was expressed by Aristotle. In reply, let me say two things.

First, the position does not imply that we should abandon the search for more and better reasons and more critical insight in the domain of practical decision. It is just that our capacity to resolve conflicts in particular cases may extend beyond our capacity to enunciate general principles that explain those resolutions. Perhaps we are working with general principles unconsciously, and can discover them by codifying our decisions and particular intuitions. But this is not necessary either for the operation or for the development of judgment. Second, the search for general principles in ethics, or other aspects of practical reasoning, is more likely to be successful if systematic theories restrict themselves to one aspect of the subject—one component of rational motivation—than if they try to be comprehensive.

To look for a single general theory of how to decide the right thing to do is like looking for a single theory of how to decide what to believe. Such progress as we have made in the systematic justification and criticism of beliefs has not come mostly from general principles of reasoning but from the understanding of particular areas, marked out by the different sciences, by history, by mathematics. These vary in exactness, and large areas of belief are left out of the scope of any theory. These must be governed by common sense and ordinary, prescientific reasoning. Such reasoning must also be used where the results of various more systematic methods bear on the matter at hand, but no one of them determines a conclusion. In civil engineering problems, for example, the solution depends both on physical factors capable of precise calculation and behavioral or psychological factors that are not. Obviously one should use exact principles and methods to deal with those aspects of a problem for which they are available, but sometimes there are other aspects as well, and one must resist the temptation either to ignore them or to treat them by exact methods to which they are not susceptible.

We are familiar with this fragmentation of understanding and method when it comes to belief, but we tend to resist it in the case of decision. Yet it is as irrational to despair of systematic ethics because one cannot find a completely general account of what should be done as it would be to give up scientific research because there is no general method of arriving at true beliefs. I am not saying that ethics is a science, only that the relation betwen ethical theory and practical decisions is analogous to the relation between scientific theory and beliefs about particular things or events in the world.

In both areas, some problems are much purer than others, that is, their solutions are more completely determined by factors that admit of precise understanding. Sometimes the only significant factor in a practical decision is personal obligation, or general utility, and then one's reasoning can be confined to that (however precisely it may be understood). Sometimes a process of decision is artificially insulated against the influence of more than one type of factor. This is not always a good thing, but sometimes it is. The example I have in mind is the judicial process, which carefully excludes, or tries to exclude, considerations of utility and personal commitment, and limits itself to claims of right. Since the systematic recognition of such claims is very important (and also tends over the long run not to conflict unacceptably with other values), it is worth isolating these factors for special treatment. As a result, legal argument has been one of the areas of real progress in the understanding of a special aspect of practical reason. Systematic theory and the search for general principles and methods may succeed elsewhere if we accept a fragmentary approach. Utilitarian theory, for example, has a great deal to contribute if it is not required to account for everything. Utility is an extremely important factor in decisions, particularly in public policy, and philosophical work on its definition, the coordination problems arising in the design of institutions to promote utility, its connections with preference, with equality, and with efficiency, can have an impact on such decisions.

This and other areas can be the scene of progress even if none of them aspires to the status of a general and complete theory of right and wrong. There will never be such a theory, in my view, since the role of judgment in resolving conflicts and applying disparate claims and considerations to real life is indispensable. Two dangers can be avoided if this idea of noncomprehensive systematization is kept in mind. One is the danger of romantic defeatism, which abandons rational theory because it inevitably leaves many problems unsolved. The other is the danger of exclusionary over-rationalization, which bars as irrelevant or empty all considerations that cannot be brought within the scope of a general system admitting explicitly defensible conclusions. This yields skewed results by counting only measurable or otherwise precisely describable factors, even when others are in fact relevant. The alternative is to recognize that the legitimate grounds of decision are extremely various and understood to different degrees. This has both theoretical and practical implications.

On the theoretical side, I have said that progress in particular areas of ethics and value theory need not wait for the discovery of a general foundation (even if there is such a thing). This is recognized by many philosophers and has recently been urged by John Rawls, who claims not only that the pursuit of substantive moral theory, for example the theory of justice, can proceed independently of views about the foundation of ethics, but that until substantive theory is further developed, the search for foundations may be premature.[2]

This seems too strong, but it is certainly true of any field that one need not make progress at the most fundamental level to make progress at all. Chemistry went through great developments during the century before its basis in atomic physics came to be understood. Mendelian genetics was developed long before any understanding of the molecular basis of heredity. At present, progress in psychology must be pursued to a great extent independently of any idea about its basis in the operation of the brain. It may be that all psychological phenomena are ultimately explainable in terms of the theory of the central nervous system, but our present understanding of that system is too meager to permit us even to look for a way to close the gap.

The corresponding theoretical division in ethics need not be so extreme. We can continue to work on the foundations while exploring the superstructure, and the two pursuits should reinforce each other. I myself do not believe that all value rests on a single foundation or can be combined into a unified system, because different types of values represent the development and articulation of different points of view, all of which combine to produce decisions. Ethics is unlike physics, which represents one point of view, that which apprehends the spatiotemporal properties of the universe described in mathematical terms. Even in this case, where it is reasonable to seek a unified theory of all physical phenomena, it is also possible to understand a great deal about more particular aspects of the physical universe—gravitation, mechanics, electromagnetic fields, radiation, nuclear forces—without having such a theory.

But ethics is more like understanding or knowledge in general than it is like physics. Just as our understanding of the world involves various points of view—among which the austere viewpoint of physics is the most powerfully developed and one of the most important—so values come from a number of viewpoints, some more personal than others, which cannot be reduced to a common denominator any more than history, psychology, philology, and economics can be reduced to physics. Just as the types of understanding available to us are distinct, even though they must all coexist and cooperate in our minds, so the types of value that move us are disparate, even though they must cooperate as well as they can in determining what we do.

With regard to practical implications, it seems to me that the fragmentation of effort and of results that is theoretically to be expected in the domain of value has implications for the strategy to be used in applying these results to practical decisions, especially questions of public policy. The lack of a general theory of value should not be an obstacle to the employment of those areas of understanding that do exist; and we know more than is generally appreciated. The lack of a general theory leads too easily to a false dichotomy: either fall back entirely on the unsystematic intuitive judgment of whoever has to make a decision, or else cook up a unified but artificial system like cost-benefit analysis,[3] which will grind out decisions on any problem presented to it. (Such systems may be useful if their claims and scope of operation are less ambitious.) What is needed instead is a mixed strategy, combining systematic results where these are applicable with less systematic judgment to fill in the gaps.

However, this requires the development of an approach to decisions that will use available ethical understanding where it is relevant. Such an approach is now being sought by different groups working in applied ethics, with what success we shall not know for some time. I want to suggest that the fragmentation of value provides a rationale for a particular way of looking at the task, and an indication of what needs to be done.

What we need most is a method of breaking up or analyzing practical problems to say what evaluative principles apply, and how. This is not a method of decision. Perhaps in special cases it would yield a decision, but more usually it would simply indicate the points at which different kinds of ethical considerations needed to be introduced to supply the basis for a responsible and intelligent decision. This component approach to problems is familiar enough in connection with other disciplines. It is expected that important policy decisions may depend on economic factors, political factors, ecological factors, medical safety, scientific progress, technological advantages, military security, and other concerns. Advice on all these matters can be obtained by responsible officials if there is anyone available whose job it is to think about them. In some cases well-established disciplines are involved. Their practitioners may vary widely in understanding of the subject, and on many issues they will disagree with one another. But even to be exposed to these controversies (about inflation or nuclear power safety or recombinant DNA risk) is better than hearing nothing at all. Moreover it is important that within most serious disciplines there is agreement about what is controversial and what is not. Anyone with an important decision to make, whether he is a legislator or a cabinet officer or a department official, can get advice on different aspects of the problem from people who have thought much more than he has about each of those aspects, and know what others have said about it. The division of disciplines and a consensus about what dimensions of a problem have to be considered are very useful in bringing together the problems and such expertise as there is.

We need a comparable consensus about what important ethical and evaluative questions have to be considered if a policy decision is to be made responsibly. This is not the same thing as a consensus in ethics. It means only that there are certain aspects of any problem that most people who work in ethics and value theory would agree should be considered, and can be professionally considered in such a way that whoever is going to make the decision will be exposed to the relevant ideas currently available. Sometimes the best ideas will not be very good, or they will include diametrically opposed views; but this is true everywhere, not just in ethics.

It might be suggested that the best approach would be to emulate the legal system by setting up an advocacy procedure before a kind of court whose job would be to render decisions on ethically loaded policy questions. (The recent proposal of a science court shows the attractions of the legal model: its nondemocratic character has great intellectual appeal.) But I think the actual situation is too fluid for anything like that. Values are relevant to policy in too many ways, and in combination with too many other kinds of knowledge and opinion, to be treatable in this manner. Although some legal decisions are very difficult, courts are designed to decide clear, narrowly defined questions to which a relatively limited set of arguments and reasons is relevant. (Think of the function of a judge in striking material from the record or refusing to admit certain data or testimony in evidence: such restrictions do not in general apply to legislative or administrative deliberations.) Most practical issues are much messier than this, and their ethical dimensions are much more complex. One needs a method of insuring that where relevant understanding exists, it is made available, and where there is an aspect of the problem that no one understands very well, this is understood too.

I have not devised such a method, but clearly it would have to provide that factors considered should include, among others, the following: economic, political, and personal liberty, equality, equity, privacy, procedural fairness, intellectual and aesthetic development, community, general utility, desert, avoidance of arbitrariness, acceptance of risk, the interests of future generations, the weight to be given to interests of other states or countries. There is much to be said about each of these. The method would have to be more organized to be useful, but a general position on the ways in which ethics is relevant to policy could probably be agreed on by a wide range of ethical theorists, from relativists to utilitarians to Kantians. Radical disagreement about the basis of ethics is compatible with substantial agreement about what the important factors are in real life. If this consensus, which I believe already exists among ethical theorists, were to gain wider acceptance among the public and those who make policy, then the extensive but fragmented understanding that we possess in this area could be put to better use than it is now. It would then be more difficult simply to ignore certain questions, and even if the ethical considerations, once offered, were disregarded or rejected, the reasons or absence of reasons for such rejection would become part of the basis for any decision made. There is a modicum of power even in being able to state a *prima facie* case.

This conception of the role of moral theory also implies an answer to the question of its relation to politics, and other methods of decision. Ethics is not being recommended as a decision procedure, but as an essential resource for making decisions, just as physics, economics, and demography are. In fundamental constitutional decisions of the Supreme Court, one branch of ethics plays a central role in a process that takes precedence over the usual methods of political and administrative decision. But for most of the questions that need deciding, ethical considerations are multiple, complex, often cloudy, and mixed up with many others. They need to be considered in a systematic way, but in most cases a reasonable decision can be reached only by sound judgment, informed as well as possible by the best arguments that any relevant disciplines have to offer.

Notes

1. See Gilbert Harman, "Practical Reasoning," *Review of Metaphysics,* XXIX (1976), 432–63.

2. John Rawls, *A Theory of Justice* (Cambridge: Harvard University Press, 1971), pp. 51–60. See also "The Independence of Moral Theory," *Proceedings and Addresses of the American Philosophical Association* (1974–75) pp. 5–22.

3. See Lawrence Tribe, "Policy Science: Analysis or Ideology?," *Philosophy & Public Affairs,* II, number 1 (Fall, 1972), 66–110.

10

Moral Dilemmas and Consistency

Ruth Barcan Marcus

I want to argue that the existence of moral dilemmas, even where the dilemmas arise from a categorical principle or principles, need not and usually does not signify that there is some inconsistency (in a sense to be explained) in the set of principles, duties, and other moral directives under which we define our obligations either individually or socially. I want also to argue that, on the given interpretation, consistency of moral principles or rules does not entail that moral dilemmas are resolvable in the sense that acting with good reasons in accordance with one horn of the dilemma erases the original obligation with respect to the other. The force of this latter claim is not simply to indicate an intractable fact about the human condition and the inevitability of guilt. The point to be made is that, although dilemmas are not settled without residue, the recognition of their reality has a dynamic force. It motivates us to arrange our lives and institutions with a view to avoiding such conflicts. It is the underpinning for a second-order regulative principle: that as rational agents with some control of our lives and institutions, we ought

From *The Journal of Philosophy* 77 (1980): 121–36. Reprinted with the permission of *The Journal of Philosophy* and the author.

to conduct our lives and arrange our institutions so as to minimize predicaments of moral conflict.

I

Moral dilemmas have usually been presented as predicaments for individuals. Plato, for example, describes a case in which the return of a cache of arms has been promised to a man who, intent on mayhem, comes to claim them. Principles of promise-keeping and benevolence generate conflict. One does not lack for examples. It is safe to say that most individuals for whom moral principles figure in practical reasoning have confronted dilemmas, even though these more commonplace dilemmas may lack the poignancy and tragic proportions of those featured in biblical, mythological, and dramatic literature. In the one-person case there are principles in accordance with which one ought to do x and one ought to do y, where doing y requires that one refrain from doing x; i.e., one ought to do not-x. For the present rough-grained discussion, the one-person case may be seen as an instance of the n-person case under the assumption of shared principles. Antigone's sororal (and religious) obligations conflict with Creon's obligations to keep his word and preserve the peace. Antigone is obliged to arrange for the burial of Polyneices; Creon is obliged to prevent it. Under generality of principles they are each obliged to respect the obligations of the other.

It has been suggested that moral dilemmas, on their face, seem to reflect some kind of inconsistency in the principles from which they derive. It has also been supposed that such conflicts are products of a plurality of principles and that a single-principled moral system does not generate dilemmas.

In the introduction to the *Metaphysics of Morals* Kant[1] says, "Because however duty and obligation are in general concepts that express the objective practical necessity of certain actions . . . it follows . . . that a conflict of duties and obligations is inconceivable *(obligationes non colliduntur)*." More recently John Lemmon,[2] citing a familiar instance of dilemma, says, "It may be argued that our being faced with this moral situation merely reflects an implicit inconsistency in our existing moral code; we are forced, if we are to remain both moral and logical, by the situation to restore consistency to our code by adding exception clauses to our present principles or by giving priority to one principle over another, or by some such device. The situation is as it is in mathematics: there, if an inconsistency is revealed by derivation, we are compelled to modify our axioms; here, if an inconsistency is revealed in application, we are forced to revise our principles." Donald Davidson,[3] also citing examples of conflict, says, "But then unless we take the line that moral principles *cannot* conflict in application to a case, we must give up the concept of the nature of practical reason we have so far been assuming. For how can premises, all of which are true (or acceptable), entail a contradiction? It is astonishing that in contemporary moral philosophy this problem has received little attention and no satisfactory treatment."

The notion of inconsistency which views dilemmas as evidence for inconsistency seems to be something like the following. We have to begin with a set of one or more moral principles which we will call a *moral code*. To count as a principle in such a code, a precept must be of a certain generality; that is, it cannot be tied to specific individuals at particular times or places, except that on any occasion of use it takes the time of that occasion as a zero coordinate. The present rough-grained discussion does not require that a point be made of the distinction between categorical moral principles and conditional moral principles, which impose obligations upon persons in virtue of some condition, such as that of being a parent, or a promise-maker or contractee. For our purposes we may think of categorical principles as imposing obligations in virtue of one's being a person and a member of a moral community.

In the conduct of our lives, actual circumstances may arise in which a code mandates a course of action. Sometimes, as in dilemmas, incompatible actions x and y are mandated; that is, the doing of x precludes the doing of y; y may in fact be the action of refraining from doing x. The underlying view that takes dilemmas as evidence of inconsistency is that a code is consistent if it applies without conflict to all actual—or, more strongly—to all possible cases. Those who see a code as the foundation of moral reasoning and adopt such a view of consistency argue that the puzzle of dilemmas can be resolved by elaboration of the code: by hedging principles with exception clauses, or establishing a rank ordering of principles, or both, or a procedure of assigning weights, or some combination of these. We need not go into the question of whether exception clauses can be assimilated to priority rankings, or priority rankings to weight assignments. In any case, there is some credibility in such solutions, since they fit some of the moral facts. In the question of whether to return the cache of arms, it is clear (except perhaps to an unregenerate Kantian) that the principle requiring that the promise be kept is overridden by the principle requiring that we protect human lives. Dilemmas, it is concluded, are merely apparent and not real. For, with a complete set of rules and priorities or a complete set of riders laying out circumstances in which a principle does not apply, in each case one of the obligations will be vitiated. What is incredible in such solutions is the supposition that we could arrive at a complete set of rules, priorities, or qualifications which would, in every possible case, unequivocally mandate a single course of action; that where, on any occasion, doing *x* conflicts with doing *y,* the rules with qualifications or priorities will yield better clear reasons for doing one than for doing the other.

The foregoing approach to the problem of moral conflict—ethical formalism—attempts to dispel the reality of dilemmas by expanding or elaborating on the code. An alternative solution, that of moral intuitionism, denies that it is possible to arrive at an elaboration of a set of principles which will apply to all particular circumstances. W. D. Ross,[4] for example, recognizes that estimates of the stringency of different *prima facie* principles can sometimes be made, but argues that no general universally applicable rules for such rankings can be laid down. However, the moral intuitionists *also* dispute the reality of moral dilemmas. Their claim is that moral codes are only guides; they are not the only and ultimate ground of decision making. *Prima facie* principles play an important heuristic role in our deliberations, but not as a set of principles that can tell us how we ought to act in all particular circumstances. That ultimate determination is a matter of intuition, albeit rational intuition. Moral dilemmas are *prima facie,* not real conflicts. In apparent dilemmas there *is* always a correct choice among the conflicting options; it is only that, and here Ross quotes Aristotle, "the decision rests with perception." For Ross, those who are puzzled by moral dilemmas have failed to see that the problem is epistemological and not ontological, or real. Faced with a dilemma generated by *prima facie* principles, *uncertainty* is increased as to whether, in choosing *x* over *y,* we have in fact done the right thing. As Ross puts it, "Our judgments about our actual duty in concrete situations, have none of the certainty that attaches to our recognition of general principles of duty. . . . Where a possible act is seen to have two characteristics in virtue of one of which it is *prima facie* right and in virtue of the other *prima facie* wrong we are well aware that we are not certain whether we ought or ought not to do it. Whether we do it or not we are taking a moral risk" (30) [pp. 95–96 in this volume]. For Ross, as well as the formalist, it is only that we may be uncertain of the right way. To say that dilemma is evidence of inconsistency is to confuse inconsistency with uncertainty. There *is* only one right way to go, and hence no problem of inconsistency.

There are, as we see, points of agreement between the formalist and the intuitionist as here described. Both claim that the appearance of dilemma and inconsistency flows from *prima facie* principles and that dilemmas can be resolved by supplementation. They differ in the nature of the supplementation.[5] They further agree that it is the multiplicity of principles which generates the *prima facie* conflicts; that if there were one rule or principle or maxim, there would be no conflicts. Quite apart from the unreasonableness of the belief that we can arrive ultimately at a single moral principle, such proposed single principles have played a major role in moral philosophy, Kant's categorical imperative and various versions of the principle of utility being primary examples. Setting aside the casuistic logical claim that a single principle can always be derived by conjunction from a multiplicity, it can be seen that the single-principle solution is mistaken. There is always the analogue of Buridan's ass. Under the single principle of promise-keeping, I might make two promises in all good faith and reason that they will not conflict, but then they do, as a result of circumstances that were unpredictable and beyond my control. All other considerations may balance out. The lives of identical twins are in jeopardy, and, through force of circumstances, I am in a position to save only one. Make the situation as symmetrical as you please. A single-principled framework is not necessarily unlike the code with qualifications or priority rule, in that it would appear that, however strong our wills and complete our knowledge, we might be faced with a moral choice in which there are no moral grounds for favoring doing x over y.

Kant imagined that he had provided a single-principled framework from which all maxims flowed. But Kantian ethics is notably deficient in coping with dilemmas. Kant seems to claim that they don't really arise, and we are provided with no moral grounds for their resolution.

It is true that unregenerate act utilitarianism is a plausible candidate for dilemma-free principle or conjunction of principles, but not because it can be framed as a single principle. It is rather that attribution of rightness or wrongness to certain kinds of acts per se is ruled out whether they be acts of promise-keeping or promise-breaking, acts of trust or betrayal, of respect or contempt. One might, following Moore, call such attributes "non-natural kinds," and they enter into all examples of moral dilemmas. The attribute of having maximal utility as usually understood is not such an attribute. For to the unregenerate utilitarian it is not features of an act per se which make it right. The only thing to be counted is certain consequences, and, for any given action, one can imagine possible circumstances, possible worlds if you like, in each of which the action will be assigned different values—depending on different outcomes in those worlds. In the unlikely cases where in fact two conflicting courses of action have the same utility, it is open to the act utilitarian to adopt a procedure for deciding, such as tossing a coin.

In suggesting that, in all examples of dilemma, we are dealing with attributions of rightness per se independent of consequences is not to say that principles of utility do not enter into moral dilemmas. It is only that such conflicts will emerge in conjunction with nonutilitarian principles. Indeed, such conflicts are perhaps the most frequently debated examples, but not, as we have seen, the only ones. I would like to claim that it is a better fit with the moral facts that all dilemmas are real, even where the reasons for doing x outweigh, and in whatever degree, the reasons for doing y. That is, wherever circumstances are such that an obligation to do x and an obligation to do y cannot as a matter of circumstance be fulfilled, the obligations to do each are not erased, even though they are unfulfillable. Mitigating circumstances may provide an explanation, an excuse, or a defense, but I want to claim that this is not the same as denying one of the obligations altogether.

We have seen that one of the motives for denying the reality of moral dilemmas is to preserve, on some notion of consistency, the consistency of our moral reasoning. But other not unrelated reasons have been advanced for denying their reality which have to do with the notion of guilt. If an agent ought to do x, then he is guilty if he fails to do it. But if, however strong his character and however good his will and intentions, meeting other equally weighted or overriding obligations precludes his doing x, then we cannot assign guilt, and, if we cannot, then it is incoherent to suppose that there is an obligation. Attendant feelings of the agent are seen as mistaken or misplaced.

That argument has been rejected by Bas van Fraassen[6] on the ground that normative claims about when we ought to assign guilt are not part of the analysis of the concept of guilt, for if it were, such doctrines as that of "original sin" would be rendered incoherent. The Old Testament assigns guilt to three or four generations of descendants of those who worship false gods. Or consider the burden of guilt borne by all the descendants of the house of Atreus, or, more recently, the readiness of many Germans to assume a burden of guilt for the past actions of others. There are analogous converse cases, as in the assumption of guilt by parents for actions of adult children. Having presented the argument, I am not wholly persuaded that a strong case can be made for the coherence of such doctrines. However, the situation faced by agents in moral dilemmas is not parallel. Where moral conflict occurs, there is a genuine sense in which both what is done and what fails to be done are, before the actual choice among irreconcilable alternatives, within the agent's range of options. But, as the saying goes—and it is not incoherent—you are damned if you do and you are damned if you don't.

I will return to the question of the reality of moral dilemmas, but first let me propose a definition of consistency for a moral code which is compatible with that claim.

II

Consistency, as defined for a set of meaningful sentences or propositions, is a property that such a set has if it is possible for all of the members of the set to be true, in the sense that contradiction would not be a logical consequence of supposing that each member of the set is true. On that definition "grass is white" and "snow is green" compose a consistent set although false to the facts. There is a possible set of circumstances in which those sentences are true, i.e., where snow is green and grass is white. Analogously we can define a set of rules as consistent if there is some possible world in which they are all obeyable in all circumstances in *that* world. (Note that I have said "obeyable" rather than "obeyed" for I want to allow for the partition of cases where a rule-governed action fails to be done between those cases where the failure is a personal failure of the agent—an imperfect will in Kant's terms—and those cases where "external" circumstances prevent the agent from meeting conflicting obligations. To define consistency relative to a kingdom of ends, a deontically perfect world in which all actions that ought to be done are done, would be too strong; for that would require both perfection of will *and* the absence of circumstances that generate moral conflict.) In such a world, where all rules are obeyable, persons intent on mayhem have not been promised or do not simultaneously seek the return of a cache of arms. Sororal obligations such as those of Antigone do not conflict with obligations to preserve peace, and so on. Agents may still fail to fulfill obligations.

Consider, for example, a silly two-person card game. (This is the partial analogue of a two-person dilemma. One can contrive silly games of solitaire for the one-person dilemma.) In the two-person game the deck is shuffled and divided equally, face down between two players. Players turn up top cards on each play until the cards are played out. Two rules are in force: black cards trump red cards, and high cards (ace high) trump lower-valued cards without attention to color. Where no rule applies, e.g., two red deuces, there is indifference and the players proceed. We could define the winner as the player with the largest number of tricks when the cards are played out. There is an inclination to call such a set of rules inconsistent. For suppose the pair turned up is a red ace and a black deuce; who trumps? This is not a case of rule indifference as in a pair of red deuces. Rather, two rules apply, and both cannot be satisfied. But, on the definition here proposed, the rules are consistent in that there are possible circumstances where, in the course of playing the game, the dilemma would not arise and the game would proceed to a conclusion. It is possible that the cards be so distributed that, when a black card is paired with a red card, the black card happens to be of equal or higher value. Of course, with shuffling, the likelihood of dilemma-free circumstances is small. But we could have invented a similar game where the likelihood of proceeding to a conclusion without dilemma is greater. Indeed a game might be so complex that its being dilemmatic under any circumstances is very small and may not even be known to the players.[7] On the proposed definition, rules are consistent if there are possible circumstances in which no conflict will emerge. By extension, a set of rules is inconsistent if there are *no* circumstances, no possible world, in which all the rules are satisfiable.[8]

A pair of offending rules which generates inconsistency as *here* defined provides *no* guide to action under any circumstance. Choices are thwarted whatever the contingencies. Well, a critic might say, you have made a trivial logical point. What pragmatic difference is there between the inconsistent set of rules and a set, like those of the game described above, where there is a likelihood of irresolvable dilemma? A code is, after all, supposed to guide action. If it allows for conflicts without resolution, if it tells us in some circumstances that we ought to do x and we ought to do y even though x and y are incompatible in those circumstances, that is tantamount to telling us that we ought to do x and we ought to refrain from doing x and similarly for y. The code has failed us as a guide. If it is not inconsistent, then it is surely deficient, and, like the dilemma-provoking game, in need of repair.

But the logical point is not trivial, for there are crucial disanalogies between games and the conduct of our lives. It is part of the canon of the family of games of chance like the game described, that the cards must be shuffled. The distribution of the cards must be "left to chance." To stack the deck, like loading the dice, is to cheat. But, presumably, the moral principles we subscribe to are, whatever their justification, not justified merely in terms of some canon for games. Granted, they must be guides to action and hence not totally defeasible. But consistency in our sense is surely only a necessary but not a sufficient condition for a set of moral rules. Presumably, moral principles have some ground; we adopt principles when we have reasons to believe that they serve to guide us in right action. Our interest is not merely in having a playable game whatever the accidental circumstances, but in doing the right thing to the extent that it is possible. We want to maximize the likelihood that in all circumstances we can act in accordance with each of our rules. To that end, our alternative as moral agents, individually and collectively, as contrasted with the card-game players, is to try to stack the deck so that dilemmas do not arise.

Given the complexity of our lives and the imperfection of our knowledge, the occasions of dilemma cannot always be foreseen or predicted. In playing games, when we are faced with a conflict of rules we abandon the game or invent new playable rules; dissimilarly, in the conduct of our lives we do not abandon action, and there may be no justification for making new rules to fit. We proceed with choices as best we can. Priority rules and the like assist us in those choices and in making the best of predicaments. But, if we do make the best of a predicament, and make a choice, to claim that one of the conflicting obligations has thereby been erased is to claim that it would be mistaken to feel guilt or remorse about having failed to act according to that obligation. So the agent would be said to believe falsely that he is guilty, since his obligation was vitiated and his feelings are inappropriate. But that is false to the facts. Even where priorities are clear and overriding and even though the burden of guilt may be appropriately small, explanations and excuses are in order. But in such tragic cases as that described by Jean-Paul Sartre[9] where the choice to be made by the agent is between abandoning a wholly dependent mother and not becoming a freedom fighter, it is inadequate to insist that feelings of guilt about the rejected alternative are mistaken and that assumption of guilt is inappropriate. Nor is it puritanical zeal which insists on the reality of dilemmas and the appropriateness of the attendant feelings. For dilemmas, when they occur, are data of a kind. They are to be taken into account in the future conduct of our lives. If we are to avoid dilemmas we must be motivated to do so. In the absence of associated feelings, motivation to stack the deck, to arrange our lives and institutions so as to minimize or avoid dilemma is tempered or blunted.

Consider, for example, the controversies surrounding nonspontaneous abortion. Philosophers are often criticized for inventing bizarre examples and counterexamples to make a philosophical point. But no contrived example can equal the complexity and the puzzles generated by the actual circumstances of foetal conception, parturition, and ultimate birth of a human being. We have an organism, internal to and parasitic upon a human being, hidden from view but relentlessly developing into a human being, which at some stage of development can live, with nurture, outside of its host. There are arguments that recognize competing claims: the right to life of the foetus (at some stage) versus the right of someone to determine what happens to his body. Arguments that justify choosing the mother over the foetus (or vice versa) where their survival is in competition. Arguments in which foetuses that are defective are balanced against the welfare of others. Arguments in which the claims to survival of others will be said to override survival of the foetus under conditions of great scarcity. There are even arguments that deny *prima facie* conflicts altogether on some metaphysical grounds, such as that the foetus is not a human being or a person until quickening, or until it has recognizable human features, or until its life can be sustained external to its host, or until birth, or until after birth when it has interacted with other persons. Various combinations of such arguments are proposed in which the resolution of a dilemma is seen as more uncertain, the more proximate the foetus is to whatever is defined as being human or being a person. What all the arguments seem to share is the assumption that there is, despite uncertainty, a resolution without residue; that there is a correct set of metaphysical claims, principles, and priority rankings of principles which will justify the choice. Then, given the belief that one choice is justified, assignment of guilt relative to the overridden alternative is seen as inappropriate, and feelings of guilt or pangs of conscience are viewed as, at best, sentimental. But as one tries to unravel the tangle of arguments, it is clear that to insist that there is in every case a solution without residue is false to the moral facts.

John Rawls,[10] in his analysis of moral sentiments, says that it is an essential characteristic of a moral feeling that an agent, in explaining the feeling, "invokes a moral concept and its associated principle. His (the agent's) account of his feelings makes reference to an acknowledged right or wrong." Where those ingredients are absent, as, for example, in the case of someone of stern religious background who claims to feel guilty when attending the theater although he no longer believes it is wrong, Rawls wants to say that such a person has certain sensations of uneasiness and the like which resemble those he has when he feels guilty, but, since he is not apologetic for his behavior, does not resolve to absent himself from the theater, does not agree that negative sanctions are deserved, he experiences not a feeling of guilt, but only something like it. Indeed, it is the feeling which needs to be explained; it is not the action which needs to be excused. For, says Rawls, in his discussion of moral feelings and sentiments, "When plagued by feelings of guilt . . . a person wishes to act properly in the future and strives to modify his conduct accordingly. He is inclined to admit what he has done, to acknowledge and accept reproofs and penalties." Guilt qua feeling is here defined not only in terms of sensations but also in terms of the agent's disposition to acknowledge, to have wishes and make resolutions about future actions, to accept certain outcomes, and the like. Where an agent acknowledges conflicting obligations, unlike the theater-goer who acknowledges no obligation, there is sufficient overlap with dilemma-free cases of moral failure to warrant describing the associated feelings where present as guilt, and where absent as appropriate to an agent with moral sensibility. Granted that, unlike agents who fail to meet their obligations simpliciter, the agent who was confronted with a dilemma may finally act on the best available reasons. Still, with respect to the rejected alternative he acknowledges a wrong in that he recognizes that it was within his power to do otherwise. He may be apologetic and inclined to explain and make excuses. He may sometimes be inclined to accept external reproofs and penalties. Not perhaps those which would be a consequence of a simple failure to meet an obligation but rather like the legal cases in which mitigating circumstances evoke a lesser penalty—or reproof.[11]

Even if, as Rawls supposes, or hopes (but as seems to me most unlikely), a complete set of rules and priorities were possible which on rational grounds would provide a basis for choosing among competing claims in all cases of moral conflict that actually arise, it is incorrect to suppose that the feeling evoked on such occasions, if it is evoked, only resembles guilt, and that it is inappropriate on such occasions to ascribe guilt. *Legal* ascriptions of guilt require sanctions beyond the pangs of conscience and self-imposed reproofs. In the absence of clear external sanctions, legal guilt is normally not ascribable. But that is one of the many distinctions between the legal and the moral.

Most important, an agent in a predicament of conflict will also "wish to act properly in the future and strive to modify his actions accordingly." He will strive to arrange his own life and encourage social arrangements that would prevent, to the extent that it is possible, future conflicts from arising. To deny the appropriateness or correctness of ascriptions of guilt is to weaken the impulse to make such arrangements.[12]

III

I have argued that the consistency of a set of moral rules, even in the absence of a complete set of priority rules, is not incompatible with the reality of moral dilemmas. It would appear, however, that at least some versions of the principle "'ought' implies 'can'" are being denied; for dilemmas are circumstances where, for a pair of obligations, if one is satisfied then the other cannot be. There is, of course, a range of interpretations of the precept resulting from the various interpretations of 'ought', 'can', and 'implies'. Some philosophers who recognize the reality of dilemmas have rejected the precept that "'ought' implies 'can'"; some have accepted it.[13] If we interpret the 'can' of the precept as "having the ability in this world to bring about," then, as indicated above, in a moral dilemma, 'ought' *does* imply 'can' for *each* of the conflicting obligations, *before* either one is met. And after an agent has chosen one of the alternatives, there is still something which he ought to have done and could have done and which he did not do. 'Can', like 'possible', designates a modality that cannot always be factored out of a conjunction. Just as 'possible P and possible Q' does not imply 'possible both P and Q', so 'A can do x and A can do y' does not imply 'A can do both x and y'. If the precept "'ought' implies 'can'" is to be preserved, it must also be maintained that 'ought' designates a modality that cannot be factored out of a conjunction. From 'A ought to do x' and 'A ought to do y' it does not follow that 'A ought to do x and y'. Such a claim is of course a departure from familiar systems of deontic logic.

The analysis of consistency and dilemmas advanced in this paper suggests a second-order principle which relates 'ought' and 'can' and which provides a plausible gloss of the Kantian principle "Act so that thou canst will thy maxim to become a universal law of nature." As Kant understood laws of nature, they are, taken together, universally and jointly applicable in all particular circumstances. It is such a second-order principle that has been violated when we knowingly make conflicting promises. It is such a second-order principle that has, for example, been violated when someone knowingly and avoidably conducts himself in such a way that he is confronted with a choice between the life of a foetus, the right to determine what happens to one's body, and benefits to others. To will maxims to become universal laws we must will the means, and among those means are the conditions for their compatibility. One ought to act in such a way that, if one ought to do x and one ought to do y, then one can do both x and y. But the second-order principle is regulative. This second-order 'ought' does *not* imply 'can'.[14] There is no reason to suppose, this being the actual world, that we can, individually or collectively, however holy our wills or rational our strategies, succeed in foreseeing and wholly avoiding such conflict. It is not merely failure of will, or failure of reason, which thwarts moral maxims from becoming universal laws. It is the contingencies of this world.

IV

Where does that leave us? I have argued that all dilemmas are real in a sense I hope has been made explicit. Also that there is no reason to suppose on considerations of consistency that there *must* be principles which, on moral grounds, will provide a sufficient ordering for deciding all cases. But, it may be argued, when confronted with what are *apparently* symmetrical choices undecidable on moral grounds, agents do, finally, choose. That is sometimes understood as a way in which, given good will, an agent makes explicit the rules under which he acts. It is the way an agent discovers a priority principle under which he orders his actions. I should like to question that claim.

A frequently quoted remark of E. M. Forster[15] is "if I had to choose between betraying my country and betraying my friend, I hope I should have the courage to betray my country." One could of course read that as if Forster had made manifest some priority rule: that certain obligations to friends override obligations to nation. But consider a remark of A. B. Worster, "if I had to choose between betraying my country and betraying my friend, I hope I should have the courage to betray my friend." Both recognize a dilemma, and one can read Worster as subscribing to a different priority rule and, to that extent, a different set of rules from Forster's. But is that the only alternative? Suppose Forster had said that, morally, Worster's position is as valid as his own. That there was no moral reason for generalizing his own choice to all. That there was disagreement between them not about moral principles but rather about the kind of persons they wished to be and the kind of lives they wished to lead. Forster may not want Worster for a friend; a certain possibility of intimacy may be closed to them which perhaps Forster requires in a friend. Worster may see in Forster a sensibility that he does not admire. But there is no reason to suppose that such appraisals are or must be moral appraisals. Not all questions of value are moral questions, and it may be that not all moral dilemmas are resolvable by principles for which moral justification can be given.

Notes

This paper was written during my tenure as a Fellow at the Center for the Advanced Study of the Behavioral Sciences. I am grateful to Robert Stalnaker for his illuminating comments. A version of the paper was delivered on January 17, 1980, at Wayne State University as the Gail Stine Memorial Lecture.

1. Immanuel Kant, *The Metaphysical Elements of Justice:* Part I of the *Metaphysics of Morals,* translated by John Ladd (Indianapolis: Bobbs-Merrill, 1965), p. 24 [p. 39 in this volume; the translation is different].

2. "Deontic Logic and the Logic of Imperatives," *Logique et Analyse,* VIII, 29 (April 1965): 39–61. Lemmon originally presented his paper at a symposium of the Western Division meeting of the American Philosophical Association in May 1964. My unpublished comments on that occasion contain some of the ideas here presented.

3. "How is Weakness of the Will Possible?," in Joel Feinberg, ed., *Moral Concepts* (New York: Oxford, 1970), p. 105.

4. *The Right and the Good* (New York: Oxford, 1930), p. 41 [pp. 98–99 in this volume].

5. For the formalist, priority rankings (like Rawls's lexical ordering), or weights permitting some computation, or qualifications of principles to take care of all problematic cases, are supposed possible. For the intuitionist it is intuitive "seeing" in each case which supplements *prima facie* principles.

6. "Values and the Heart's Command," *Journal of Philosophy,* LXX, 1 (Jan. 11, 1973): 5–19 [chapter 7 in this volume]. Van Fraassen makes the point that such a claim would make *the* doctrine of "original sin" incoherent. As I see it, there are at least three interesting doctrines, two of them very likely true, which could qualify as doctrines of original sin. One of them, which I call "inherited guilt," is the doctrine that some of the wrongful actions of some persons are such that other persons, usually those with some special connection to the original sinners, are also judged to be sinners; their feelings of guilt are appropriate, their punishment "deserved," and so on. Such is the case described in Exodus and Deuteronomy here mentioned. A second notion of original sin is to be found in the account of the Fall. Here it is suggested that, however happy our living arrangements, however maximal the welfare state, we will each of us succumb to some temptation. There is universality of sin because of universality of weakness of will, but specific sins are neither inherited by nor bequeathed to others.

A third candidate supposes the reality and inevitability, for each of us, of moral dilemma. Here we do not inherit the sins of others, nor need we be weak of will. The circumstances of the world conspire against us. However perfect our will, the contingencies are such that situations arise where, if we are to follow one right course of action, we will be unable to follow another.

7. There is a question whether, given such rules, the "game" is properly described as a game. Wittgenstein says "Let us suppose that the game [which I have invented] is such that whoever begins can always win by a particular simple trick. But this has not been realized;—so it is a game. Now someone draws our attention to it—and it stops being a game." *Remarks on the Foundations of Mathematics,* ed., G. H. von Wright *et al.,* translated by G. E. M. Anscombe (Oxford: Blackwell, 1956), II 78, p. 100e. Wittgen-

stein is pointing to that canon of a game which requires that both players have some opportunity to win. The canon that rules out dilemmatic rules is that the game must be playable to a conclusion. (I am beholden to Robert Fogelin for reminding me of this quotation.)

8. Bernard Williams, in *Problems of the Self* (New York: Cambridge, 1977), chapters 11 [chapter 6 in this volume] and 12, also recognizes that the source of some apparent inconsistencies in imperatives and rules is to be located in the contingency of their simultaneous inapplicability on a given occasion.

9. Sartre in "Existentialism Is a Humanism" describes a case where a student is faced with a decision between joining the Free French forces and remaining with his mother. He is her only surviving son and her only consolation. Sartre's advice was that "No rule of general morality can show you what you ought to do." His claim is that in such circumstances "nothing remains but to trust our instincts." But what is "trust" here? Does our action reveal to us that we subscribe to a priority principle or that in the absence of some resolving principles we may just as well follow our inclination? In any case to describe our feelings about the rejected alternative as "regret" seems inadequate. See Walter Kaufmann, ed., *Existentialism from Dostoevsky to Sartre* (New York: Meridian, 1956), pp. 295–98.

10. *A Theory of Justice* (Cambridge, Mass.: Harvard, 1971), pp. 481–83. Rawls's claim is that such sensations, to be properly describable as "guilt feelings" and not something resembling such feelings, must occur in the broader context of beliefs, strivings, acknowledgements, and readiness to accept outcomes, and cannot be detached from that context. He rejects the possibility that there are such "pure" sensations that can occur independent of the broader context. This is partially, perhaps, an empirical claim about identifying sameness of feeling. The theater-goer might claim that he does feel guilty because he has the same feeling he has when he acknowledges that he is guilty, that what remains is to give an account of when such feelings of guilt are justified. Still, Rawls's analysis seems to me to be a better account.

11. To insist that "regret" is appropriate rather than "guilt" or "remorse" is false to the facts. It seems inappropriate, for example, to describe as "regret" the common feelings of guilt that women have in cases of abortion even where they believe (perhaps mistakenly) that there was moral justification in such an undertaking.

12. Bernard Williams ["Politics and Moral Character," in Stuart Hampshire, ed., *Public and Private Morality* (New York: Cambridge, 1978), pp. 54–74] discusses the question in the context of politics and the predicament of "dirty hands." He argues that, where moral ends of politics justify someone in public life lying, or misleading, or using others, "the moral disagreeableness of these acts is not merely cancelled." In particular, we would not want, as our politicians, those "practical politicians" for whom the disagreeableness does not arise.

13. For example, John Lemmon, in "Moral Dilemmas," *Philosophical Review*, LXXI, 2 (April 1962): 139–58, p. 150 [p. 114 in this volume], rejects the principle that 'ought' implies 'can'. Van Fraassen, *op. cit.,* pp. 12–13 [p. 146 in this volume], accepts it, as does Bernard Williams seemingly in *Problems of the Self, op. cit.,* pp. 179–84 [pp. 129–34 in this volume]. Van Fraassen and Williams see that such acceptance requires modification of the principle of factoring for the deontic "ought." There are other received principles of deontic logic which will have to be rejected, but they will be discussed in a subsequent paper. It should also be noted that, in "Ethical Consistency" and "Consistency and Realism" in *Problems of the Self,* Williams also articulates the contingent source of dilemmas and argues for their "reality."

14. See note 13. The reader is reminded that, on the present analysis, 'ought' is indexical in the sense that applications of principles on given occasions project into the future. They concern bringing something about.

15. *Two Cheers for Democracy* (London: E. Arnold, 1939).

11

Moral Conflicts

R. M. Hare

1

I shall be introducing a distinction between two levels of moral thinking. The distinction is not original; it occurs already in Plato and Aristotle. The seeds of it are to be found in Plato's distinction between knowledge and right opinion. . . . This is the basis of his division between the classes in his *Republic,* and the educations proper to them.[1] It reappears in Aristotle's distinction between right motivation and practical wisdom, virtues of character and of intellect, the 'that' and the 'why'. It was made use of by the classical utilitarians in answering their critics.[2] But it may be doubted whether its immense importance has yet been realized. [It] is hardly an exaggeration to say that more confusion is caused, both in theoretical ethics and in practical moral issues, by the neglect of this distinction than by any other factor.

I shall be using the distinction, first in order to shed light on some disputes in metaethics which have recently troubled us, and secondly in order to defend a version of utilitarianism against an extremely common type of objection which would not be made by anybody who understood the distinction. But the uses made of the distinction [that I shall make here] by no means exhaust its usefulness; in particular, the philosophy of education, as I shall be briefly hinting, can gain a great deal of illumination from it.

I shall be calling the two levels the intuitive and the critical, following my practice in recent papers.[3] Earlier I called them level 1 and level 2[4]; but it is clearer to have more self-explanatory names for them. To these two levels of moral thinking has to be added a third level, the metaethical, at which we are operating when we discuss the meanings of the moral words and the logic of moral reasoning. The intuitive and critical levels of thinking are both, unlike the metaethical, concerned with moral questions of substance; but they handle them in different ways, each appropriate to the different circumstances in which, and purposes for which, the thinking is done.

The way in which I shall seek to explain the two levels is by discussing the problem of moral conflicts. By this I mean situations in which we seem to have conflicting duties. The views held by moral philosophers about conflicts of duties are an extremely good diagnostic of the comprehensiveness and penetration of their thought about morality; superficiality is perhaps more quickly revealed by what is said about this problem than in any other way. Those who say, roundly, that there can just be irresoluble conflicts of duties are always those who have confined their thinking about morality to the intuitive level. At this level the conflicts are indeed irresoluble; but at the critical level there is a requirement that we resolve the conflict, unless we are to confess that our thinking has been incomplete. We are not thinking critically if we just say 'There is a conflict of duties; I ought to do *A,* and I ought to do *B,* and I can't do both'. But at the intuitive level it is perfectly permissible to say this. The critical level is that at which the minister was operating who put a placard on the 'wayside pulpit' outside his church in Yorkshire (as reported to me by Mr. Anthony Kenny) saying 'If you have conflicting duties, one of them isn't your duty'.

It can readily be agreed, however, that it is sometimes the case that a person thinks that he ought to do *A,* and also thinks that he ought to do *B,* but cannot do both. For example, he may have made a promise, and circumstances may have intervened, by no fault of his, such that he has an urgent duty to perform which precludes his fulfilling the promise. To start with the kind of trivial example which used to be favoured by intuitionist philosophers: I have promised to take my children for a picnic on the river at Oxford, and then a lifelong friend turns up from Australia and is in Oxford for the afternoon, and wants to be shown round the colleges with his wife. Clearly I ought to show them the colleges, and clearly I ought to keep my promise to my children. Not only do I think these things, but in some sense I am clearly right.

If I am in this dilemma, I may decide on reflection, and in spite of thinking, and *going on* thinking, that I ought to keep my promise to my children, that what I ought, all things considered, to do is to take my friend round the colleges; and this involves a decision that *in some sense* I ought not to take my children for their picnic, because it would preclude my doing what, all things considered, I ought to do, namely take my friend round the colleges. In general, it seems that we think that, if I ought to do *A,* and doing *B* would preclude my doing *A,* I ought not to do *B.* I rely here on our linguistic and logical, not our moral, intuitions. And if this is so, it looks as if I both ought, and ought not, to take my children for their picnic. And it is a small step from this to concluding that I ought to both take them for their picnic and not take them. This, obviously, runs counter to the claim that 'ought' implies 'can'; for I cannot both take them and not take them. And, even if we do not make the move from saying that I ought to do *A* to saying that I ought not to do *B* because it would preclude doing *A,* we can still get a breach of the dictum that 'ought' implies 'can' by going from the statement that I ought to do *A* and ought to do *B* to the statement that I ought to do *A* and *B;* for I cannot, in this imagined case, do both *A* and *B.*

Faced with these difficulties, some have taken the heroic course of saying that from the proposition that I ought to do *A* and ought to do *B* it does not follow that I ought to do *A* and *B.* And others might take the equally heroic course of saying that it is possible for it to be the case that I ought to do *A* and ought not to do *A,* but that this does not entail that I ought to both do *A* and not do *A.* Others, again, might abandon the dictum that 'ought' implies 'can'. There is indeed some case on other grounds for limiting the application of this dictum.[5] But even its complete abandonment would remove only one, and that not the most awkward, of the difficulties presented by cases of moral conflict.

It is not very helpful to try to sort out these difficulties by some relatively minor tinkering with the calculus of deontic logic. The linguistic and logical intuitions which give rise to them are all right so far as they go. We may suspect, rather, that there is some ambiguity, or at least difference of use, affecting 'ought' as it occurs in different contexts; and this is the suggestion which I shall in effect be making. What is required is an understanding of the two levels of moral thinking and of the different ways that 'ought' is used in each of them. I say this, because, if we have this understanding, the difficulties disappear, and because there are many independent reasons for taking this step.

2

It will be illuminating at this point to take a look at the phenomena of moral conflict and ask why philosophers, and indeed most of us, are so certain that we sometimes ought to do two things of which we cannot do both. One clue is to be found in an argument sometimes advanced. It is said that, whichever of the things we do, we shall, if we are morally upright people, experience *remorse,* and that this is inseparable from thinking that we ought not to have done what we did. If, it is said, we just stopped having a duty to do one of the things because of the duty to do the other (as on the wayside pulpit) whence comes the remorse?

This might be questioned. It might be said that, though *regret* is in place (for my children, after all, have had to miss their picnic, and that is a pity), remorse is not; just because it implies the thought that I ought not to have done what I did (a thought which I do not have; I do not reject my decision that, all things considered, I ought to do what I did), it is irrational to feel remorse. On this account, the philosophers in question have just confused remorse with regret. Or they might be said to have confused either of these things with a third thing, which we might call, following Sir David Ross,[6] *compunction.* This feeling is not easily distinguishable from fear; and it afflicts us during or before the doing of an act, unlike remorse, which only occurs afterwards. Compunction is also not normally so strong a feeling as remorse; but, like remorse, it *can* be irrational. It would be a very hardened intuitionist who maintained that we never have these feelings in situations in which it is wholly absurd to have them. So perhaps, on the occasion I have been describing, regret would be in place (just as it would be if the picnic or the tour of the colleges had to be cancelled because of the weather), but remorse and compunction are not.

We may, however, feel a lingering unease at this reply. Would not the man who could break his promise to his children without a twinge of compunction be, not a better, because more rational, person, but a morally worse one than most of us who are afflicted in this way? It is time we came to more serious examples. The following kind of case is often cited. A person falls overboard from a ship in a wartime convoy; if the master of the ship leaves his place in the convoy to pick him up, he puts the ship and all on board at risk from submarine attack; if he does not, the person will drown. In the film *The Cruel Sea*,[7] a somewhat similar case occurs; the commander of a corvette is faced with a situation in which if he does not drop depth charges the enemy submarine will get away to sink more ships and kill more people; but if he does drop them he will kill the survivors in the water. In fact he drops them, and is depicted in the film as suffering anguish of mind. And we should think worse of him if he did not. The former case is perhaps the better one to take, because it avoids some irrelevant features of the second; it too would cause anguish. Although we might feel tempted to say that this anguish is just extreme regret and not remorse, because the master had decided that he ought to leave the person to drown, and remorse would imply his thinking that he ought not to have left him to drown, which, *ex hypothesi,* he does not think—all the same we may also feel that there is some residuum of *moral* sentiment in his state of mind which is not mere nonmoral regret. This, at any rate, is the source of the intuitionist view that we are discussing, that there are irresoluble moral conflicts.

Reverting to the case of promising: let us suppose that I have been well brought up; I shall then think, let us say, that one ought not to break promises. And suppose that I get into a situation in which I think that, in the circumstances, I ought to break a promise. I cannot then just abolish my past good upbringing and its effects; nor should I wish to. If I have been well brought up, I shall, when I break the promise, experience this feeling of compunction (no doubt 'remorse' would be too strong a word in this case), which could certainly be described, *in a sense,* as 'thinking that I ought not to be doing what I am doing, namely breaking a promise'. This is even clearer in the case of lying. Suppose that I have been brought up to think that one ought not to tell lies, as most of us have been. And suppose that I get into a situation in which I decide that I ought, in the circumstances, to tell a lie. It does not follow in the least that I shall be able to tell the lie without compunction. That is how lie detectors work (on people who have been to this minimal extent well brought up). Even if I do not blush, something happens to the electrical properties of my skin. And for my part, I am very glad that this happens to my skin; for if it did not I should be a morally worse educated person.

When we bring up our children, as both Plato and Aristotle[8] agreed, one of the things we are trying to do is to cause them to have reactions of this kind: 'to like and dislike the things they ought to like and dislike'. It is not by any means the whole of moral education, although some people speak as if it were; indeed, since one can do this sort of thing even to a dog, it may not be part of a typically human moral education at all, or at any rate not the distinctive part. People who think that it is the whole of moral education, and call it 'teaching children the difference between right and wrong', do not have my support, though they are very numerous (which explains why intuitionism is such a popular view in moral philosophy). But there is no doubt that most of us have, during our upbringing, acquired these sentiments, and not much doubt that this is, on the whole, a good thing, for reasons which will shortly become evident.

I recently visited Prague to talk to some philosophers there. If, when I was crossing into Czechoslovakia, the officials had asked me the purpose of my visit, I should certainly have told them a lie, because if they had known they would most probably have expelled me, as they have some of my colleagues, these visits being frowned upon by the Czech government. And, just as certainly, I should have felt, not merely fear of being found out and getting into trouble, but a feeling of *guilt* at telling the lie (although I should have been in no doubt that I ought to tell it). A lie detector would certainly have exposed my deception. And I should be a morally worse man if I were not affected in that way. So then, if I felt guilty, it looks as if there is a sense of 'thinking that I ought' in which I could have been correctly described as 'thinking that I ought not to be telling the lie'. Feeling guilty is inseparable from, in this sense, thinking that I ought not. Only those with some philosophical sophistication are able to distinguish them even in thought.[9] Yet in another sense I should certainly have been thinking that I ought to tell the lie.

3

Nobody who actually uses moral language in his practical life will be content with a mere dismissal of the paradox that we can feel guilty for doing what we think we ought to do. It will not do to say "There just are situations in which, whatever you do, you will be doing what you ought not, i.e., doing wrong'. There are, it is true, some people who like there to be what they call 'tragic situations'; the world would be less enjoyable without them, for the rest of us: we could have much less fun writing and reading novels and watching films, in which such situations are a much sought after ingredient. The trouble is that, if one is enough of an ethical descriptivist[10] to make the view that there are such situations tenable, they stop being tragic. If, that is to say, it just is the case that both the acts open to a person have the moral property of wrongness, one of their many descriptive properties, why should he be troubled by that? What makes the situation tragic is that he is using moral thinking to help him to decide what he ought to *do;* and when he does this with no more enlightenment than that provided by those 'absolutist' thinkers who believe in very simple and utterly inviolable principles, it leads to an impasse. He is like a rat in an insoluble maze, and that is tragic. But the very tragedy of it may make more humane philosophers look for an alternative to the theory which puts him there. In such a conflict between intuitions, it is time to call in reason.

However, it is not immediately obvious that a rational approach to such predicaments requires the complete separation of levels of moral thinking which I am going to postulate. Let us try to do without it. One expedient might be to seek to qualify or modify one of the principles which seem to apply to the situation, introducing some saving clause which will make the two principles no longer conflict, even *per accidens* in this situation. (Two principles conflict *per accidens* when, although the conjunction of the two is not self-contradictory, which would be a conflict per se, they cannot, as things happen to be, both be complied with.) In order to display the logic of the conflict briefly, I shall have to be formal and use letters for the features of situations which figure in the conflicting principles.

Let us then suppose that I start off with principles that one ought never to do an act which is F (some property of acts, e.g., being an instance of promise-breaking), and that one ought never to do an act which is G (some other property), and that I find myself in a situation in which I cannot avoid doing either an F act or a G act. And suppose that I decide, all things considered, that I ought to do the G act, to escape the (worse) F act. We might then try supposing that, as a result of this piece of moral thinking (however it was done), one of my moral principles has *changed.* Instead of reading 'One ought never to do an act which is G', it now reads 'One ought never to do an act which is G, unless it is necessary in order to avoid an act which is F'. This principle, then, which was only ten words long (counting 'G' as one word, though in fact it stands for many words), has now attained a length of twenty-three words. This is important, because the philosophers I have in mind seem often to be speaking as if moral principles ought to be very simple and general, and if one were required to quantify simplicity or generality, one crude way of doing it would be in terms of the number of words in a principle; and they do sometimes talk as if no decent moral principle would contain more than, say, a dozen words.

But obviously this lengthening process has only just begun. By the time we have been in, or even considered without actually being in them, a few such dilemmas, we shall be getting very long principles indeed. Very early on we shall get principles like 'One ought never to do an act which is G, except that one may when it is necessary in order to avoid an act which is F, and the act is also H; but if the act is not H, one may not' (43 words). For reasons of space I shall not pursue this process further,[11] but shall assume that it will soon get out of hand, if what we are after are simple principles. And that certainly seems to be what the philosophers I am thinking of are after, because those are the principles which they like appealing to in their examples, and the making of exceptions to which they resist.

Another expedient would be to say that neither of the two rival principles is qualified, but that I adopt a superior or second-order principle which says, for example, 'The principle that one ought never to do an act which is G ought to be obeyed in all cases except where obeying it involves a breach of the principle that one ought never to do an act which is F; in such cases one ought to break it', plus, of course, additional complications to deal with the 'H' factor just mentioned. Even without these additional complications the new principle contains 49 words. It is obvious that this expedient, though it yields a longer and more cumbrous principle, is no different in substance from the first one. It is also obvious that any attempt to set up what is called a hierarchy of principles, telling us which is to override which, and when, will, if it is to do justice to the moral judgements which we actually make in cases of conflict, soon become extremely complicated; for the instructions for operating the hierarchy will mostly be in the form of the 49-word principle just mentioned. There is the additional problem, brought out by the *'H'* factor in the 43-word principle of the preceding para- that we are unlikely to be able always to put principles in the same order of priority; shall we not want to say that we ought to tell lies to avoid giving pain in *some* circumstances, but that we ought to give pain in order to avoid telling lies in *other* circumstances (for example when the pain is not great)?

Some may wish to introduce God on his traditional machine to ensure that there are no real conflicts of duties even in a one-level system. I shall not discuss this view, because I do not think that even many Christians will believe that God has fixed things in this way, but rather that God has, and we have to a much more limited degree, the means (rational moral thinking) wherewith to resolve at the critical level conflicts which arise at the intuitive. [Elsewhere I] profess a less extravagant faith about how the world has to be ordered if morality is to be viable [*MT,* chapter 11, section 8].

I shall also not discuss a fourth expedient, which has been quite popular: that of saying that when two principles or two duties conflict *per accidens,* the conflict is to be settled by a judging or weighing process to determine which ought, in the particular case, to be conformed to. This is less a method than an evasion of the problem; we are not told *how* to weigh or judge the rival principles. It is, further, not a one-level but a two-level procedure, without any explanation of what is supposed to happen at the second level. The procedure can, however, be made viable within an adequate two-level system [see section 4 below].

I can think of no other way of registering what happens in cases where conflicts of principles are decided that is open to the user of a one-level structure of moral thinking. Since, therefore, a one-level structure seems condemned either to having no determinate procedure for settling moral conflicts, or to having principles of ever increasing complexity, we can only be content with it if we are happy with the complexity or the indeterminacy. I am not yet taking sides on the question of how simple moral principles have to be. As we shall see, it depends on the purposes for which the principles are to be used; different sorts of principles are appropriate to different roles in our moral thinking.

4

Let us then look at some of these roles and purposes. One very
important one is in *learning* [*LM,* chapter 4, section 3 (footnote);
chapter 8, section 3; chapter 10, section 3]. If principles reach more
than a certain degree of complexity, it will be impossible to formulate them verbally in sentences of manageable length; but it might
still be possible, even after that, to learn them—i.e., to come to
know them in some more Rylean sense which does not involve
being able to recite them.[12] Assuredly there are many things we
know without being able to say in words what we know. All the
same, there is a degree of complexity, higher than this, beyond
which we are unable to learn principles even in this other sense
which does not require that we be able to recite them. So principles which are to be learnt for use on subsequent occasions have to
be of a certain degree of simplicity, although the degree has been
exaggerated by some people.

In addition to this psychological reason for a limit to the complexity of principles, there is also a practical reason related to the
circumstances of their use. Situations in which we find ourselves are
not going to be minutely similar to one another. A principle which
is going to be useful as a practical guide will have to be unspecific
enough to cover a variety of situations all of which have certain
salient features in common. What is wrong about situational ethics
and certain extreme forms of existentialism (we shall see in a
moment what is right about them) is that they make impossible
what is in fact an indispensable help in coping with the world
(whether we are speaking of moral decisions or of prudential or
technical ones, which in this are similar), namely the formation in
ourselves of relatively simple reaction patterns (whose expression
in words, if they had one, would be relatively simple prescriptive
principles) which prepare us to meet new contingencies resembling
in their important features contingencies in which we have found
ourselves in the past. The same trouble afflicts the crude caricature
of act utilitarianism which is the only version of it that many philosophers seem to be acquainted with. If it were not possible to form
such dispositions, any kind of learning of behaviour would be ruled
out, and we should have to meet each new situation entirely unprepared, and perform an 'existential' choice or a cost-benefit analysis
on the spot. Let anybody who is tempted by doctrines of this kind
think what it would be like to drive a car without having *learnt* how

to drive a car, or having totally forgotten everything that one had ever learnt—to drive it, that is, deciding *ab initio* at each moment what to do with the steering wheel, brake, and other controls.

This point can also be illustrated by the practice of good chess players and by the experience of those who have designed computers to play this and other games. Even the best chess players cannot explore more than a few moves ahead. It is therefore rare for them, until late in the game, to win games by working out all the branching alternative moves up to the end of the game, i.e., by calculating which of the available moves will put them in a position which, whatever the opponent does, enables them to make a further move which will put them in a position which, whatever the opponent does, will put them in a position which . . . (after a very large number of such calculated moves) . . . enables them to checkmate the opponent. Nor can any computers so far developed complete such calculations up to the checkmate in the time allowed at chess tournaments.

The point is even clearer with games like backgammon, in which the use of dice introduces an element of chance, thereby vastly multiplying the alternative possibilities that have to be considered at each move. Backgammon is actually a better analogue of our ordinary human problems than chess; that is why its simpler analogues among children's games are such a good moral preparation for facing 'the changes and chances of this mortal life'. The designer of a computer which recently beat the world champion at backgammon makes it clear that the way he did it was not by exploiting the power of the computer to explore beyond human limits the consequences of alternative moves, but rather by programming the computer with relatively simple principles for selecting promising moves (i.e., those most likely to improve the strength of one's position, as assessed by another set of principles picking out general features of positions which are important in *those* kinds of positions) and for selecting *which* of the many alternative future courses of the game to explore more fully, and for how far. 'Unlike the best programmes for playing chess, BKG 9.8 does well more by positional judgment than by brute calculation. This means that it plays backgammon much as human experts do'.[13] Obviously judgement that can be programmed into a computer is not open to the strictures I made in [section 3 above] against 'judging or weighing'. It is governed by principles or rules; a method has been given. No doubt good military commanders and good politicians operate in the same way as the good backgammon players whom this computer emulates, and so do good moralists.

It by no means follows from this that *who wins* a game of backgammon, a battle, or an election is defined by reference to these rules of good play. The winner of a game of backgammon is the player who first bears off all his pieces in accordance with the rules of the game, not the one who follows the best strategies: Similarly in morals, the principles which we have to follow if we are to give ourselves the best chance of acting rightly are not definitive of 'the right act'; but if we wish to act rightly we shall do well, all the same, to follow them. A wise act utilitarian, unlike his caricature mentioned earlier, will agree with this, so he is not vulnerable to objections based on it.

A further reason for relying in much of our moral conduct on relatively general principles is that, if we do not, we expose ourselves to constant temptation to special pleading [*FR*, chapter 3, section 6]. In practice, especially when in haste or under stress, we may easily, being human, 'cook' our moral thinking to suit our own interest. For example, it is only too easy to persuade ourselves that the act of telling a lie in order to get ourselves out of a hole does a great deal of good to ourselves at relatively small cost to anybody else; whereas, in fact, if we view the situation impartially, the indirect costs are much greater than the total gains. It is partly to avoid such cooking that we have intuitions or dispositions firmly built into our characters and motivations.

The term 'rules of thumb' is sometimes used in this connexion, but should be avoided as thoroughly misleading. I regret having once used it myself [*LM*, chapter 4, section 4]. Some philosophers use it in a quite different way from engineers, gunners, navigators and the like, whose expression it really is, and in whose use a rule of thumb is a mere time- and thought-saving device, the breach of which, unlike the breach of the moral principles we are discussing, excites no compunction. A much better expression is '*prima facie* principles'. Such principles express '*prima* facie duties',[14] and, although formally speaking they are just universal prescriptions, are associated, owing to our upbringing, with very firm and deep dispositions and feelings. Any attempt to drive a wedge between the principles and the feelings will falsify the facts about our intuitive thinking. *Having* the principles, in the usual sense of the word, is having the disposition to experience the feelings, though it is not, as some intuitionists would have us believe, incompatible with submitting the principles to critical thought when that is appropriate and safe.

There are, then, both practical and psychological reasons for having relatively simple principles of action if we are to learn to behave either morally or skilfully or with prudence. The situational ethicists have rejected this obvious truth because they have grasped another obvious truth which they think to be incompatible with it, though it will seem to be so only to someone who has failed to make the distinction between the two levels of moral thinking which I shall be postulating. The situations in which we find ourselves are like one another, sometimes, in some important respects, but not like one another in all respects; and the differences may be important too. 'No two situations and no two people are ever exactly like each other': this will be recognized as one of the battle cries of the school of thought that I am speaking of.

It follows from this that, although the relatively simple principles that are used at the intuitive level are necessary for human moral thinking, they are not sufficient. Since any new situation will be unlike any previous situation in *some* respects, the question immediately arises whether the differences are relevant to its appraisal, moral or other. If they are relevant, the principles which we have learnt in dealing with past situations may not be appropriate to the new one. So the further question arises of how we are to decide whether they are appropriate. The question obtrudes itself most in cases where there is a conflict between the principles we have learnt—i.e., where, as things contingently are, we cannot obey them both. But if it arises in those cases, it can arise in any case, and it is mere intellectual sloth to pretend otherwise.

5

The most fundamental objection to the one-level account of moral thinking called intuitionism is that it yields no way of answering such a question. The intuitive level of moral thinking certainly exists and is (humanly speaking) an essential part of the whole structure; but however well equipped we are with these relatively simple, *prima facie,* intuitive principles or dispositions, we are bound to find ourselves in situations in which they conflict and in which, therefore, some other, nonintuitive kind of thinking is called for, to resolve the conflict. The intuitions which give rise to the conflict are the product of our upbringings and past experience of decision-making. They are not self-justifying; we can always ask whether the upbringing was the best we could have, or whether the past decisions were the right ones, or, even if so, whether the principles then formed should be applied to a new situation, or, if they cannot all be applied, *which* should be applied. To use intuition itself to answer such questions is a viciously circular procedure; if the dispositions formed by our upbringing are called into question, we cannot appeal to them to settle the question.

What will settle the question is a type of thinking which makes no appeal to intuitions other than linguistic. I stress that in this other kind of thinking, which I am calling *critical* thinking, no moral intuitions of substance can be appealed to. It proceeds in accordance with canons established by philosophical logic and thus based on linguistic intuitions only. To introduce substantial moral intuitions at the critical level would be to incorporate in critical thinking the very same weakness which it was designed to remedy. A philosopher will not be content with the intuitive props on which most moral philosophers rely, if he wishes his work to last.

Critical thinking consists in making a choice under the constraints imposed by the logical properties of the moral concepts and by the nonmoral facts, and by nothing else. This choice is what I used to call a decision of principle [*LM,* chapter 4]. But the principles involved here are of a different kind from the *prima facie* principles considered so far. Since some people have been misled by the term 'principle', I have asked myself whether I should avoid it altogether; but I have in the end retained it in order to mark an important logical similarity between the two kinds of principles. Both are universal prescriptions; the difference lies in the generality specificity dimension. To explain this: a *prima facie* principle has, for reasons I have just given, in order to fulfil its function, to be relatively simple and general (i.e., unspecific). But a principle of the kind used in critical thinking (let us call it a critical moral principle) can be of unlimited specificity.

There is not space here to explain at length the difference between universality and generality. It is very important, and it is a pity that these words are still often used as if there were no distinction.[15] My own terminology in *LM* [chapter 10, section 3] was slovenly. Briefly, generality is the opposite of specificity, whereas universality is compatible with specificity, and means merely the logical property of being governed by a universal quantifier and not containing individual constants. The two principles 'Never kill people' and 'Never kill people except in self-defence or in cases of adultery or judicial execution' are both equally universal, but the first is more general (less specific) than the second. Critical principles and *prima facie* principles, then, are both universal prescriptions; but whereas the former can be, and for their purposes have to be, highly specific, the latter can be, and for *their* purposes have to be, relatively general. Just *how* general they should be will depend on the circumstances and temperaments of individuals. I have discussed this question more fully elsewhere.[16]

6

Let us, after these preliminaries, return to our conflict-situation, in which two *prima facie* principles require two incompatible actions. This will be because one of the principles picks out certain features of the situation as relevant (e.g., that a promise has been made), and the other picks out certain others (e.g., that the failure to show my friend the colleges would bitterly disappoint him). The problem is to determine which of these principles should be applied to yield a prescription for this specific situation. The method to be employed in critical thinking can only be briefly sketched here [a full account is given in *MT,* part 2].

Notice first that in theory it is not necessary (though it usually is in practice), in order to describe a situation fully, or as fully as we need, to mention individuals. We can describe it in universal terms, including in the description the alternative actions that are open and their respective consequences. We can, however, without omitting any *descriptive* information, omit all individual references, so that the description will apply equally to any precisely similar situation involving precisely similar people, places, etc. Individual references preceded by 'like', 'similar', and equivalent expressions are exempted from this ban; they can be treated as universal in the required sense.[17]

The thesis of universalizability requires that if we make any moral judgement about this situation, we must be prepared to make it about any of the other precisely similar situations. Note that these do not have to be *actual* situations; they can be precisely similar logically possible hypothetical situations [*FR,* chapter 6, section 4; *MT,* chapter 6, section 4]. Therefore the battle cry referred to earlier,

'No two situations and no two people are ever exactly like each other', is not relevant in this part of moral thinking, and the thought that it is relevant is due only to confusion with other parts. What critical thinking has to do is to find a moral judgement which the thinker is prepared to make about this conflict-situation and is also prepared to make about all the other similar situations. Since these will include situations in which he occupies, respectively, the positions of all the other parties in the actual situation, no judgement will be acceptable to him which does not do the best, all in all, for all the parties. Thus the logical apparatus of universal prescriptivism, if we understand what we are saying when we make moral judgements, will lead us in critical thinking (without relying on any substantial moral intuitions) to make judgements which are the same as a careful act utilitarian would make. We see here how the utilitarians and Kant get synthesized.[18]

Much of the controversy about act utilitarianism and rule utilitarianism has been conducted in terms which ignore the difference between the critical and intuitive levels of moral thinking. Once the levels are distinguished, a form of utilitarianism becomes available which combines the merits of both varieties.[19] The conformity (for the most part) to received opinion which rule utilitarianism is designed to provide is provided by the *prima facie* principles used at the intuitive level; but critical moral thinking, which selects these principles and adjudicates between them in cases of conflict, is act utilitarian in that, in considering cases, actual or hypothetical, it can be completely specific, leaving out no feature of an act that could be alleged to be relevant. But since, although quite specific, it takes no cognizance of individual identities, it is also rule utilitarian in that version of the rule utilitarian doctrine which allows its rules to be of unlimited specificity, and which therefore is in effect not distinguishable from act utilitarianism.[20] The two kinds of utilitarianism, therefore, can coexist at their respective levels: the critical thinker considers cases in an act utilitarian or specific rule utilitarian way, and on the basis of these he selects, as I shall shortly be explaining, general *prima facie* principles for use, in a general rule utilitarian way, at the intuitive level.

7

We have next to ask, what the relation is between the two levels of moral thinking, and how we know when to think at one level, and when at the other. Let us be clear, first of all, that critical and intuitive moral thinking are not *rival* procedures, as much of the dispute between utilitarians and intuitionists seems to presuppose. They are elements in a common structure, each with its part to play. But how are they related?

Let us consider two extreme cases of people, or beings, one of whom would use *only* critical moral thinking and the other *only* intuitive. First, consider a being with superhuman powers of thought, superhuman knowledge and no human weaknesses. I am going to call him the archangel.[21] This 'ideal observer' or 'ideal prescriber'[22] resembles the 'clairvoyant' of *LM* [chapter 4, section 2] in his powers of prediction but adds to these the other superhuman qualities just mentioned. He will need to use only critical thinking. When presented with a novel situation, he will be able at once to scan all its properties, including the consequences of alternative actions, and frame a universal principle (perhaps a highly specific one) which he can accept for action in that situation, no matter what role he himself were to occupy in it. Lacking, among other human weaknesses, that of partiality to self, he will act on that principle, if it bids him act. The same will apply to other partialities (e.g., to our own friends and relations) which are hardly weaknesses, but which are, for reasons which I shall later explain, excluded from critical thinking, though they play a large part in intuitive thinking.[23] Such an archangel would not need intuitive thinking; everything would be done by reason in a moment of time. Nor, therefore, would he need the sound general principles, the good dispositions, the intuitions which guide the rest of us.

On the other hand, consider a person who has these human weaknesses to an extreme degree. Not only does he, like most of us, have to rely on intuitions and sound *prima facie* principles and good dispositions for most of the time; he is totally incapable of critical thinking (let alone safe or sound critical thinking) even when there is leisure for it. Such a person, if he is to have the *prima facie* principles he needs, will have to get them from other people by education or imitation. Let us call him the *prole* (after George Orwell in *1984*). Although the archangel and the prole are exaggerated versions of the top and bottom classes in Plato's *Republic,* it is far from my intention to divide up the human race into archangels and proles; we all share the characteristics of both to limited and varying degrees and at different times.

Our question then is, 'When ought we to think like archangels and when like proles?' Once we have posed the question in this way, the answer is obvious: it depends on how much each one of us, on some particular occasion or in general, resembles one or the other of these two characters. There is no philosophical answer to the question; it depends on what powers of thought and character each one of us, for the time being, thinks he possesses. We have to know ourselves in order to tell how much we can trust ourselves to play the archangel without ending up in the wrong Miltonic camp as *fallen* archangels.

One thing, however, is certain: that we cannot all of us, all the time, behave like proles (as the intuitionists would have us do) if there is to be a system of *prima facie* principles at all. For the selection of *prima facie* principles, and for the resolution of conflicts between them, critical thinking is necessary. If we do not think that men can do it, we shall have to invoke a Butlerian God to do it for us, and reveal the results through our consciences. But how then would we distinguish between the voice of God and the voices of our nursemaids (if we had them)?

8

I have, then, sidestepped the issue of when we should engage in these two kinds of thinking; it is not a philosophical question. The other question however is; for unless we can say how the two kinds of thinking are related to each other, we shall not have given a complete account of the structure of moral thinking.

Aristotle, in a famous metaphor, says that the relation of the intellect to the character (which is what we have been talking about in other words) has to be a paternal one: in so far as a man's motives and dispositions are rational, it is because they 'listen to reason as to a father' (1103^{a}3). Because intuitive moral thinking cannot be self-supporting, whereas critical thinking can be and is, the latter is epistemologically prior. *If* we were archangels, we could by critical thinking alone decide what we ought to do on each occasion; on the other hand, if we were proles, we could not do this, at least beyond the possibility of question, by intuitive thinking.

Provided that we do not give it a 'subjectivist' or 'relativist' interpretation [*MT,* chapter 12, section 1], there is no harm in saying that the right or best way for us to live or act either in general or on a particular occasion is what the archangel would pronounce to be so if he addressed himself to the question. This is not 'subjectivist' in any bad sense, and is certainly a highly rationalist thesis, because . . . archangels, at the end of their critical thinking, will all say the same thing [*MT,* chapter 6, section 2; chapter 12, section 3], on all questions on which moral argument is possible [*FR,* chapter 8, section 1 f; section 11 below]; and so shall we, to the extent that we manage to think like archangels. Intuitive thinking has the function of yielding a working approximation to this for those of us who cannot think like archangels on a particular occasion. If we wish to ensure the greatest possible conformity to what an archangel would pronounce, we have to try to implant in ourselves and in others whom we influence a set of dispositions, motivations, intuitions, *prima facie* principles (call them what we will) which will have this effect. We are on the whole more likely to succeed in this way than by aiming to think like archangels on occasions when we have neither the time nor the capacity for it. The *prima facie* principles themselves, however, have to be selected by critical thinking; if not by our own critical thinking, by that of people whom we trust to be able to do it.

Let us suppose that we are thus criticizing a proposed *prima facie* principle. What, as legislating members of this kingdom of ends,[24] do we actually think about? The principle is for use in our actual world. One thing, therefore, that we do *not* do is to call to mind the improbable or unusual cases that novelists, or philosophers with axes to grind, can dream up, and ask whether in *those* cases the outcome of inculcating the principle would be for the best. To take an analogous example from the prudential field: suppose that we are wondering whether to adopt the principle of always wearing our seat belts when driving. We have to balance the minor inconvenience of fastening our belts every time we drive against the serious harm we shall come to on those rare occasions, if any, on which we have collisions. Here the rarity of the occurrence is compensated for by the gravity of its consequences, and so we may well decide that it is right to adopt the principle. But then suppose somebody alleges, perhaps truly, that in *some* collisions the risk of injury or death is increased by wearing belts (for instance, when an unconscious driver would otherwise have been thrown clear of a vehicle which caught fire). There are people who fix their attention on such cases and use them as a reason for rejecting the rule to wear seat belts; and many people argue similarly in ethics, using the mere possibility or even mere conceivability of some unusual case, in which a principle would enjoin an obviously unacceptable action, as an argument for rejecting the principle. The method is unsound. Disregarding, for simplicity's sake, the difference in severity between injuries: if, say, in 95 percent of all collisions the risk of injury is reduced by wearing belts, and in 5 percent it is increased, it will be rational to wear them if we want to reduce our expectation of injury more than we want to avoid the inconvenience of wearing them.

To generalize: if we are criticizing *prima facie* principles, we have to look at the consequences of inculcating them in ourselves and others; and, in examining these consequences, we have to balance the size of the good and bad effects in cases which we consider against the probability or improbability of such cases occurring in our actual experience. It seems to be the case that popular morality has actually been caused to change without sufficient reason by failure to do this. It is very easy for a novelist (D. H. Lawrence for example) to depict with great verisimilitude, as if they were everyday occurrences, cases in which the acceptance by society of the traditional principle of, say, fidelity in marriage leads to unhappy results. The public is thus persuaded that the principle ought to be rejected. But in order for such a rejection to be rational, it would have to be the case, not merely that situations *can* occur or be conceived in which the results of the acceptance of the principle are not for the best, but that these situations are common enough to outweigh those others in which they are for the best. It is of course a matter for dispute what principle about fidelity in marriage would, on a more rational evaluation, be the one to adopt; but the evaluation has to be done. If it turned out that more suffering was caused by the breakdown of the marriage conventions than by their preservation, then an archangel, following our method [*MT,* chapter 6, section 1], and giving equal weight to the interests of each in his impartial critical thinking, would favour their preservation.

A similar, or complementary, mistake is often made by opponents of utilitarianism when they produce unusual examples (such as the sheriff who knows—who can say how?—that the innocence of the man whom he hangs in the general interest will never be exposed [*MT,* chapter 9, section 7]).[25] The purpose of [such examples] is to convince us that utilitarianism, when applied in these unusual situations, yields precepts which are at variance with our common intuitions. But this ought not to be surprising. Our common intuitions are sound ones, if they are, just because they yield acceptable precepts in common cases. For this reason, it is highly desirable that we should all have these intuitions and that our consciences should give us a bad time if we go against them. There-fore all well brought up people can be got to gang up against the utilitarian (if they can somehow be inhibited from any deep philo-sophical reflection) by citing some *un*common case, which is undoubtedly subsumable under a *prima facie* principle which we have all absorbed, and in which therefore we shall accept the utili-tarian precept, which requires a breach of the principle, only with the greatest repugnance.

These antiutilitarians sometimes overreach themselves. Profes-sor Bernard Williams, in his elaboration of a well-known example,[26] thinks that he can score against the utilitarians by showing that in this farfetched case they would have to prescribe the killing of one innocent man, the alternative being that he and nineteen others would die by another hand. We all have qualms about prescribing this—very naturally, because we have rightly been brought up to condemn the killing of innocent people, and also to condemn suc-cumbing to blackmail threats of this sort, and good utilitarian rea-sons can be given to justify such an upbringing. But when we come to consider what actually ought to be done in this bizarre situation, even Williams seems at least to contemplate the possibility of its being right to shoot the innocent man to save the nineteen other innocent men.[27] All he has shown is that we shall reach this conclu-sion with the greatest repugnance if we are 'decent' people; yet there is nothing to stop the utilitarian agreeing with this.[28]

9

To sum up, then, the relation between the two kinds of thinking is this. Critical thinking aims to select the best set of *prima facie* prin-ciples for use in intuitive thinking. It can also be employed when principles from the set conflict *per accidens*. Such employment may lead to the improvement of the principles themselves, but it need not; a principle may be overridden without being altered [section 12 below]. The best set is that whose acceptance yields actions, dispo-sitions, etc., most nearly approximating to those which would be chosen if we were able to use critical thinking all the time. This answer can be given in terms of acceptance-utility, if one is a utili-tarian; if one is not a utilitarian but a Kantian, one can say in effect the same thing by advocating the adoption of a set of maxims for general use whose acceptance yields actions, etc., most approximating to those which would be chosen if the categorical imperative were applied direct on each occasion by an archangel. Thus a clearheaded Kantian and a clearheaded utilitarian would find themselves in agreement, once they distinguished between the two kinds of thinking.

But besides the role of *selecting prima facie* principles, critical thinking has also the role of *resolving conflicts* between them. If the principles have been well chosen, conflicts will arise only in exceptional situations; but they will be agonizing in proportion as the principles are deeply held (as they should be). Though in general it is bad policy to question one's *prima facie* principles in situations of stress, because of the danger of 'cooking' already referred to, the conflicts we are speaking of force us to do this (hence the anguish). There can be different outcomes. In simpler cases we may 'feel sure' that some principle or some feature of a situation is *in that situation* more important than others (compare the backgammon parallel [section 4 above]). We shall then be able to sort the matter out intuitively, letting one principle override the other in this case, without recourse to critical thinking. This might well be best in the promise-breaking example with which we started. But though this intuitive sorting out may seem to offer a straw at which intuitionists can clutch, it is obvious that it will not be available in more serious conflicts.

At the other extreme, a conflict may force us to examine the *prima facie* principles themselves, and perhaps, instead of overriding one, qualify them from then on. People who have been through such crises often think differently thereafter about some fundamental moral questions—a sign that some critical thinking has been done, however inarticulate. This qualification of the principles will have brought with it a resolution of the conflict, because the principles as qualified are no longer inconsistent even *per accidens*. There is also a middle way: the person in the conflict-situation may come to be fairly sure that one or both of the principles ought to be qualified, but not be sure how, except that the qualification would allow such and such an accommodation in this particular case; he can then decide about the particular case, overriding one of the principles, and leave reflection on the principles themselves for another time when he is in a better position to do it rationally. Only when he has done it can he be sure that he was right.

Another version of this middle way is to say 'The principles, since they are in conflict, cannot be altogether relied on; I am compelled to depart from one or the other, and do not know which. So let me put the principles aside for the time being and examine carefully the particular case to see what critical thinking would say about it.' This is possible for critical thinking, in so far as humans can do it, and it is what the situational ethicists and crude act utilitarians might recommend in all cases. It is, as we have seen, a dangerous procedure; but sometimes we may be driven to it. Antiutilitarians make it their business to produce examples in which this is the only recourse, and then charge utilitarians with taking it (which is unavoidable) and with taking it lightheartedly (which is a slander [*MT*, chapter 8, section 1 ff.]). The good utilitarian will reach such decisions, but reach them with great reluctance because of his ingrained good principles; and he may agonize, and will certainly reflect, about them till he has sorted out by critical thinking, not only what he ought to have done in the particular case, but what his *prima facie* principles ought to be.

It may be said that one cannot compartmentalize one's moral thinking in the way the two-level account seems to require.[29] I can only reply by asking whether those who raise this objection have ever faced such situations. I do my own moral thinking in the way described [here and in *MT*] (not like an archangel, for I am not one, nor like a prole, but doing my best to employ critical and intuitive thinking as appropriate). In difficult situations one's intuitions, reinforced by the dispositions that go with them, pull one in different directions, and critical thinking, perhaps, in another. A person with any deep experience of such situations will have acquired some *methodological prima facie* principles which tell him when to launch into critical thinking and when not; they too would be justified by critical thinking in a cool hour. To say that it is impossible to keep intuitive and critical thinking going in the same thought process is like saying that in a battle a commander cannot at the same time be thinking of the details of tactics, the overall aim of victory, and the principles (economy of force, concentration of force, offensive action, etc.) which he has learnt when learning his trade. Good generals do it. The good general is one who wins his battles, not one who has the best *prima facie* principles; but the best *prima facie* principles are those which, on the whole, win battles.

10

We have now to [consider] the problem of differentiating moral from other evaluative judgements. [In *MT*, chapter 1,] I said that the differentia I should be using was that of overridingness; and this notion must therefore now be explained. This will give me an occasion to say something more about the problem of weakness of will, which has been such a crux for moral philosophers, especially for those of a prescriptivist persuasion like myself [*FR*, chapter 5, section 1]. It becomes much clearer when we see it as a particular case of the problem of conflicts between prescriptions, of which the moral conflicts dealt with [earlier] were another different, though analogous, case. It became obvious, after the levels of moral thinking were distinguished, how moral conflicts are to be accounted for; and much the same treatment helps with some important kinds of weakness of will. Those who insist that there can be conflicts of duties are quite right as regards the intuitive level; for there can indeed be conflicts of duties which are irresoluble *at that level,* and they can cause all the anguish that anybody desires. What these thinkers do not see is that there is another level, the critical, which exists for the purpose of resolving these conflicts, and also of so ordering our choice of moral principles that they will arise as infrequently as is consistent with the other purposes of moral thinking.

Both the problem of weakness of will, and that of conflicts of duties, arise because the ordinary man is firmly convinced that he ought to do certain things; and he is convinced of this because his intuitions, embodied in *prima facie* principles, assure him that this is so. So it is very easy for a philosopher to set up cases which will convince the ordinary man that he ought to do both of two incompatible things, or that he knows he ought not to be doing something which he is doing. The first of these cases is the one we have been dealing with; the second is one kind of weakness of will.

11

We have to examine the second case in more detail. But first let us take up the question of the definition of 'moral'. The two properties of universalizability and prescriptivity did not suffice to define the class of moral judgements, since at least some other, and indeed in a strict sense all, evaluative judgements have these properties [*FR,* chapter 2, section 8; chapter 8, section 2; *MT,* chapter 1, section 6]. In order to distinguish moral judgements within the larger genus, a differentia is required. The name I shall use for this distinguishing property is 'overridingness'. Though it is important that moral judgements, in one sense of that ambiguous word, have this property or (as we shall see) a property closely related to it, it will not play a very large part in our argument. Nor will the word 'moral' itself.

There are two reasons for this. The first is that we do not need this property in order to construct our account of moral reasoning. The canons of moral argument are based on the other two properties of 'ought' and 'must', which they share with all evaluative words. This is possible because moral judgements, though they are not confined to situations where the interests of others are affected, have their predominant use in such situations. For cases where the interests of others are not affected, I make no claim to provide canons of moral reasoning. But for cases where they are, the two properties of universalizability and prescriptivity suffice to govern the reasoning, and no other properties were appealed to in my account in *FR*. The reason why this account cannot be extended to argument about, e.g., aesthetic questions is that arguments about them do not turn on other people's interests, but only on the sensory and affective qualities of the aesthetic object; the reason why it cannot be extended to moral questions not affecting others' interests is the same: argument based on these two properties cannot get a grip on such questions [*FR*, chapter 8, section 1].

A second reason for avoiding, if we can, bringing the word 'moral' into our account of moral reasoning is that it is so ambiguous. It has not even a variety of well-defined uses, but a very vague spectrum of uses which shade into one another and are hard to distinguish. This is nothing unusual; Wittgenstein seems to have thought that words were typically like this,[30] and though he was exaggerating (dictionary-making is a feasible and useful activity), there are indeed many such words and 'moral' is one of them. 'Ought' and 'must' are less Protean; I feel safer with them, though they too have various uses. But I have regretfully decided that, although it is possible to base an account of moral reasoning on a certain distinguishable use of 'ought' and of 'must' (the universalizable prescriptive use), whose rules determine what we can and cannot say without self-contradiction when using the words, nothing so definite is possible with the word 'moral'.

However, we do need a concept which will delimit those uses of the universalizable prescriptive 'ought' and 'must' with which we, as moral philosophers, are concerned; and it seems to me that 'moral', in one of its uses, is the word, or a possible word, for that

concept. So the best policy will be to admit that the word is ambiguous and even vague, and to define a use of it which will mark out
those uses of 'ought' and 'must' in which we are primarily
interested.

12

We might suggest as a first approximation that a use of 'ought' or
'must' is a moral use in this sense if the judgement containing it is
(1) prescriptive; (2) universalizable; and (3) overriding. Before we
come to the inadequacies in this definition let us try to make the
third element more precise. I may say in passing that, if the attempt
to define 'moral' in these terms, or in a more developed version of
them, were successful, I should not mind *substituting* the expression
'overriding-prescriptive-universalizable' for the expression 'moral',
in this sense, if it were not so cumbrous. Then I could happily make
a present of the word 'moral' to those who wish to use it in other of
its meanings.

But what is it for one prescription to override another? I gave
a brief account of this notion in *FR* [chapter 9, section 3]. The
example I used was that of the conflict between the aesthetic principle that one ought not to juxtapose scarlet with magenta, and the
moral principle that one ought not to hurt one's wife's feelings, in a
case in which my wife has given me a magenta cushion to put on
my scarlet sofa in my room in college. I said that I would allow the
moral principle or judgement to override the aesthetic one; and this
means at least that, although both are prescriptive, I would think
that I ought to act, and accordingly would act, on the moral one and
thus not act on the aesthetic one. This, I said, was quite different
from *qualifying* the aesthetic one to admit of an exception in such
a case; the aesthetic judgement remains quite unqualified and is prescriptive, but is simply overridden in this case. I do not, thereafter,
hold the aesthetic principle in the form 'One ought not to juxtapose
scarlet and magenta, except when it is necessary in order to avoid
hurting one's wife's feelings'.

Note that if I were to treat the principle forbidding colour
clashes as overriding, and thus think that I ought to throw away the
cushion at whatever cost to my wife's feelings, I should be, on a
definition of 'moral' in terms of overridingness, elevating it into a
moral principle; whereas many definitions of 'moral' in terms of the

contents that principles called 'moral' can have would bar this. This shows that there are two senses of 'moral' involved, as I have already allowed; it does not show that mine is not a possible or useful sense, provided that it is distinguished from others.

To treat a principle as overriding, then, is to let it always override other principles when they conflict with it and, in the same way, let it override all other prescriptions, including nonuniversalizable ones (e.g., plain desires). Note that I say 'treat as'. It might be thought a defect in my account that I do not try to say what it is for a principle to *be* overriding, but only what it is to *treat* a principle as overriding. But this is a necessary feature of the definitions of many such terms. Suppose, for example, that we were to define in part the expression 'sign of conjunction' by saying that a person is treating a word (for example 'and') as a sign of conjunction if, whenever this word occurs between two sentences, he thinks it inconsistent to affirm the whole proposition expressed by the two sentences with 'and' between them, but deny a proposition expressed by one of the two sentences by itself. It would not be a valid objection to this definition to say that it only tells us what it is to *treat* 'and' as a sign of conjunction, and not what it is for it to *be* a sign of conjunction. If it is used in that way, it is, as so used, a sign of conjunction. And similarly, if someone treats a principle as overriding all others, but does not let any other override it, then the words he uses in expressing the principle are so used that they express for him an overriding principle.[31] This statement of the position does not involve us in the subjectivist view that one can make moral judgements true simply by adopting them [*MT,* chapter 12, section 1]. Whether he *ought* to be treating this principle as an overriding one is another question; but it is clear that he *is;* what he means when he utters the principle is something overriding. And if it were possible to define 'moral' in terms of overridingness, we could go on to say that, if the definition were satisfied by somebody's use of an expression, what he meant to express by it was a moral principle.

But, though this difficulty is not a real one, there are others which are. The account I have just sketched would make it impossible for a moral principle to be overridden by another moral principle, or by any other prescription. But both of these cases occur. There are both moral conflicts, which are resolved by allowing one moral principle to override another, and cases where, to use a somewhat revolting expression which I once heard a colleague use, we 'take a moral holiday'. Some cases of weakness of will are examples of this; and so also are such cases as that described by Austin in a famous passage,[32] where I deliberately, and not through weakness, when dining at High Table, help myself to two portions though there are only enough for one apiece. Our account so far is too simple to deal with these cases. But its inadequacies can be remedied by invoking our two-level structure of moral thinking. This will enable us to adjust our account of the word 'moral', and of the moral words, to their actual human uses [section 14 below].

13

Returning, then, to the problem of weakness of will, let us see what use can be made of the concept of overridingness, which must not be confused with that of prescriptivity. In *LM* [chapter 11, section 2], in a very preliminary look at the problem of weakness of will, I did indeed say that some of the cases of this were cases in which the moral judgement in question lacked prescriptivity. These cases certainly occur. Sometimes, when someone says 'I ought, but I am not going to', his moral judgement is of the 'So what?' variety [*MT*, chapter 4, section 2 *s.f.*]. But it would be unsubtle to suppose, and I did not even then suppose, that all cases of weakness of will were of this sort.

I mentioned, in fact, even at that early stage, two sorts of nonprescriptive moral judgements that might be involved (and the fact that there can be nonprescriptive moral judgements [*FR*, chapter 2, section 8] is another sign of the inadequacy of our first definition of 'moral'). The first was the 'inverted commas' moral judgement, implying merely that a certain act is required in order to conform to the moral standards current in society. The second was the moral judgement incorporated in our moral feelings. This notion was not very fully explained; but its connexion with what I said [earlier] about intuitive moral thinking will be obvious. There are two elements in this judgement which need to be distinguished: the judgement *that we have* the feeling, and the judgement which is the *expression* of the feeling. If we have been well brought up in the way that I have described, we shall both have a feeling of moral repugnance against lying, for example, and know that we have this feeling. These two factors are often jumbled up under the term 'moral intuition'.

In *FR* [chapter 5, section 9] I gave a list of possible types of weakness of will, and explicitly admitted that the list might not be exhaustive. I will not now repeat it. But two of the important items on it were, first, the case in which a man cannot resist the temptation to do what he thinks he ought not to do; secondly, the case in which he departs from what I called 'the rigour of pure prescriptive universality' such as would characterize 'a holy or angelic moral language'. I have nothing further to say here about the first of these cases; but what I have been saying [here] will be recognized as an expansion of the second [*FR*, chapter 5, section 5; *MT*, chapter 1, section 6].

Because we are human beings and not angels we have adopted or inherited what I called the intuitive level of moral thinking with its *prima facie* principles, backed up by powerful moral feelings, and attached to rather general characteristics of actions and situations. In our predicament, this is not vicious; we need this device, as I have amply explained. The *prima facie* principles are general in two connected senses; they are rather simple and unspecific, and they admit of exceptions, in the sense that it is possible to go on holding them while allowing that in particular cases one may break them. This possibility was not mentioned in *LM* [chapter 3, section 6], and should have been. In other words, they are overridable. Again, though in the sense in which I have been using the term they are universal (they contain no individual constants and start with a universal quantifier), in another sense they are not universal (they are not universally binding; one may make exceptions to them). It would be impossible for *prima facie* principles to fulfil their practical function unless they had these features, which may seem from the theoretical point of view to be faults. In order to be of use in moral education and character formation, they have to be to a certain degree simple and general; but if they are, then we shall encounter cases (the world being so various) in which to obey them (even if two of them did not conflict) would run counter to the prescriptions of an angelic moral thinking.

This fully explains why *prima facie* principles have to be overridable—why, that is to say, it is possible to go on holding them even when one does not obey them in a particular case. I repeat that this overridability does not mean that they are not prescriptive; *if* applied, they would require a certain action, but we just do not apply them in a certain case. Moreover, although I have so far considered cases in which one such *prima facie* moral principle is overridden in favour of another *prima facie* moral principle, it is likely that a principle which has this feature of overridability will also be open to being overridden by other, nonmoral, prescriptions, as when we take 'moral holidays'. That is why the whole problem becomes clearer when one sees the kind of conflict which we call weakness of will as just one example of conflicts between prescriptions. What happens when I decide that I ought to break a promise in order not to disappoint my Australian friend of his tour of Oxford has quite close affinities with what happens when I decide to break one in order not to disappoint my own appetites.

14

In the light of this admission that some moral principles can be overridden without ceasing to be held as moral principles, the reader will reasonably expect me to qualify the suggestion about the meaning of 'moral' made earlier [section 12 above]. We cannot simply say that someone is treating a universal prescriptive principle as a moral principle if and only if he does not let it be overridden by any other principles; for we have seen that *prima facie* moral principles can be overridden, not only by other moral principles, but by nonmoral prescriptions, without ceasing to be held as moral principles.

But the separation of levels, which is the cause of this difficulty, also provides its solution. If we think of the whole structure of moral thinking with its two levels, 'moral' can be defined, in the sense in which we are using it (as I said, not the only sense), as follows. The class of a man's moral principles consists of two subclasses: (1) those universal prescriptive principles which he does not allow to be overridden; these will all be what I called 'critical moral principles' [section 5 above], and are therefore capable of being made so specific and so adapted to particular cases that they do not need to be overridden; (2) those *prima facie* principles which, although they can be overridden, are selected in the way above described, by critical thinking, in the course of which use is made of moral principles of the first subclass. So, if we want to know whether someone is treating a principle as a moral principle, we have first to ask whether he would ever, in any circumstances, let it be overridden. If he says that he would not, then he is treating it as a moral principle. But even if he says that there are some circumstances in which he would let it be overridden, it might be a moral principle of the second subclass. We have to ask him, therefore, in that case, how he would justify his selection of this as one of his principles; and if he says that it would be on the basis of critical thinking, in that this was a principle whose general acceptance would lead to people's actions and dispositions approximating to the greatest extent to the deliverances of a perfectly conducted critical thinking (i.e., to the moral principles of the first sort that such critical thinking would arrive at), then this principle too will count as a moral principle, but of the second subclass.

It may be objected to this definition that it makes it impossible for proles and intuitionist philosophers, who know of only the intuitive level of moral thinking, to have any moral principles; for they cannot justify their 'moral principles' by appeal to critical thinking. It would be more correct to say that such people have no way of distinguishing their moral from their other principles. Can *we,* who know about the two levels, make the distinction on their behalf? We can, by saying that a principle is for them a moral principle if, either (1) it is treated by them as overriding (and such people may well so treat even *prima facie* principles, though it will put them in familiar straits if ever the principles conflict); or (2) *if* they were constrained (perhaps by such a conflict-situation) to do some critical thinking, however primitive, they would justify the principle by appeal to some higher principle treated as overriding. But it may be best simply to say that there is a difficulty, in the case of such people, in distinguishing their moral principles in the sense we are after; this is a sign of a gap in their thinking rather than ours.

It will be noticed that such a more complex definition of 'moral' brings us somewhat nearer to the point of view of some[33] who think of themselves as my opponents. They wish to insist that no purely formal definition of 'moral' can be given; it has to be defined either in terms of possible contents of moral principles, or in terms of possible reasons for or justifications of them. I am not for a moment abandoning my formalist position; but in spite of it (i.e., without making any other than formal moves) I have allowed that a principle is being treated as moral (of the second subclass) if the justification for it, in the mind of the person who holds it, is of a certain sort. And when we have worked out the implications of the method of critical thinking,[34] we shall see that the justifications which it provides will be of the same general sort as these writers are after. For well-conducted critical thought will justify the selection of *prima facie* principles on the ground that the general acceptance of them will lead to actions which do as much good, and as little harm, as possible.

This is another illustration, which can stand alongside that provided in an earlier article of mine,[35] of how purely formal moves can lead, by a more indirect route, to conclusions to which it is tempting to take a too hasty naturalistic short cut by writing some substance into the definition of 'moral' or of the moral words. How the more indirect route arrives at this destination become[s] clear in [*MT*] part 2.

Notes

1. Section 7 below and *Plato* in Past Masters series (Oxford University Press, 1982).

2. Mill, J.S. 1861. *Utilitarianism.* Chapter 5; cf., Rawls, J. 1955. "Two Concepts of Rules." *Philosophical Review* 64.

3. E.g., Hare, R.M. 1979. "Utilitarianism and the Vicarious Affects." *The Philosophy of Nicholas Rescher.* Ed. E. Sosa. Dordrecht: Reidel. 146.

4. E.g., Hare, R.M. 1976. "Ethical Theory and Utilitarianism." *Contemporary British Philosophy 4.* Ed. H.D. Lewis. Winchester, MA: Allen and Unwin. 122.

5. Hare, R.M. 1963. *Freedom and Reason.* Oxford: Oxford University Press. Chapter 4, section 1.

6. 1930. *The Right and the Good.* Oxford: Oxford University Press. 28 [p. 93 in this volume].

7. Adapted from Monsarrat, N. 1951. *The Cruel Sea.* London: Cassell.

8. *Nicomachean Ethics,* $1104^{b}12$ (references to Bekker's pages and columns are given in margins of most editions and translations).

9. Section 4 below; Hare, R.M. 1952. (rev. 1961). *The Language of Morals.* Oxford: Oxford University Press. Chapter 11, section 2.

10. Hare, R.M. 1981. *Moral Thinking: Its Levels, Method and Point.* Oxford: Oxford University Press. Chapter 4, section 1.

11. *See* Nozick, R. 1968. "Moral Complications and Moral Structures." *Natural Law Forum* 13.

12. Ryle, G. 1949. *The Concept of Mind.* London: Hutchinson. 27.

13. Berliner, H. 1980. "Computer Backgammon." *Scientific American* 242: 64

14. Ross, *op. cit.*: 19 [p. 86 in this volume].

15. Hare, R.M. 1972. "Principles." *Aristotelian Society* 72: 2.

16. *Ibid.*: 8.

17. Hare, R.M. 1955. "Universalisability." *Aristotelian Society* 55: 30. Reprinted in Hare, R.M. 1972. *Essays on the Moral Concepts*. London: Macmillan, and Berkeley: University of California Press.

18. For greater detail, see *MT*, part 2.

19. Hare, "Principles": 13 *ff.*; "Ethical Theory and Utilitarianism."

20. Hare, *FR*, chapter 7, section 6; Hare 1972. "Rules of War and Moral Reasoning," *Philosophy and Public Affairs* 1: 170. Reprinted in *War and Moral Responsibility*. Eds. M. Cohen *et al.* Princeton, NJ: Princeton University Press; Lyons, D. 1965. *Forms and Limits of Utilitarianism*. Oxford: Clarendon. Chapter 3.

21. Hare, "Ethical Theory and Utilitarianism": 124; cf., Godwin, W. 1793. *An Enquiry Concerning Political Justice*. London: Geoffrey Cumberlege, Oxford University Press. Book 2, chapter 2.

22. *FR*, chapter 6, section 4; Hare, "Principles": 18; Hare, "Rules of War and Moral Reasoning": 168.

23. Hare, "Utilitarianism and the Vicarious Affects"; Hare, *MT*, chapter 8, section 3.

24. Kant, I. 1785. *Grundlegung zur Metaphysik der Sitten (Groundwork to the Metaphysic of Morals)*. Translated as *The Moral Law* by H. J. Paton. 1948. London: Hutchinson, and New York: Barnes & Noble. 75 (pagination of second edition in margin).

25. More such examples are discussed in *MT*, chapter 8.

26. Williams, B.A.O. 1973. "A Critique of Utilitarianism." *Utilitarianism: For and Against*. J.J.C. Smart and B.A.O. Williams. Cambridge: Cambridge University Press. 98.

27. *Ibid.*: 117.

28. More of these manoeuvres are considered in *MT*, chapter 8, section 1 *ff.*

29. Williams, B.A.O. 1976. "Utilitarianism and Self-Indulgence." *Contemporary British Philosophy 4*, Ed. H.D. Lewis. London: Allen & Unwin. 230.

30. 1953. *Philosophical Investigations*. Oxford: Blackwell. Sections 65 *ff.*

31. It will make no difference to the argument whether we say that 'ought', as used in a principle, means the same whether or not the principle is treated as overriding, but that the person accepting the principle may, without change of meaning, treat it as overriding or as not overriding; or whether we say that if he treats it as overriding he is using 'ought' in a different way. I have assumed that the latter is the case.

32. 1956. "A Plea for Excuses." *Aristotelian Society* 57: 24n. Reprinted

in Austin, J.L. 1961. *Philosophical Papers.* Oxford: Oxford University Press.

33. E.g., Warnock, G.J. 1967. *Contemporary Moral Philosophy.* London: Macmillan, and New York: St. Martin's Press. 52 *ff.*

34. See *MT,* part 2.

35. *Essays on the Moral Concepts:* 92.

12

Against Moral Dilemmas

Earl Conee

Call an agent's predicament a "moral dilemma" just when the agent cannot do everything that it is morally obligatory for him to do in the situation, though he can carry out each obligation. A typical putative example of the phenomenon is a case where someone can keep each of two solemn trusted promises, but not both. Here I defend the view that no moral dilemma is possible. There is no fact of moral life that cannot be accounted for at least as well without moral dilemmas, and their possibility would cast a shroud of impenetrable obscurity over the concept of moral obligation.

In several recent discussions, philosophers have been indulgent of moral dilemmas. E. J. Lemmon[1] and Ruth Marcus[2] have asserted their existence, Bernard Williams[3] has said that they are required to make sense of certain moral sentiments and considerations, and Bas van Fraassen[4] has held that their existence is sufficiently tenable to mandate revising deontic logic to allow for them. I believe that none of this tolerance is justified. I will examine their principal reasons for it below, to see how well we can do without any moral dilemmas.

From *The Philosophical Review* 91 (1982): 87–97. Reprinted with the permission of *The Philosophical Review* and the author.

First, why do I think we should deny the very possibility of a moral dilemma? The basis for my principal objection was supplied by van Fraassen as he argued for making room in deontic logic for dilemmas.[5] If we accept the alleged cases as genuine, then we must reconcile ourselves to this result: actions are possible that are *both* absolutely, unconditionally, and not merely *prima facie* morally obligatory, *and* absolutely, unconditionally, and not merely *prima facie* morally impermissible. That, I submit, is absolutely, unconditionally, and not merely *prima facie* incredible. Yet the reasoning to that consequence is simple and secure. If there are moral dilemmas, then some truths have this form:

$$(1)\ O(A)\ \&\ O(\sim A)$$

For example, many an agent has made equally firm commitments to do something and to refrain from doing it. Obligation is related to permission as follows:

$$(2)\ O(A) \equiv \sim P(\sim A)$$

From (1) and (2) by propositional logic alone we can infer:

$$(3)\ O(\sim A)\ \&\ \sim P(\sim A)$$

Again I emphasize that the topic is unconditional, non-*prima facie* moral obligation. On that interpretation of "O" in (3), (3) cannot be true.

Fortunately, all available defenses of (1) and its ilk are objectionable. Let us begin to consider them. E. J. Lemmon asserts principles that would land us in many moral dilemmas. He writes:

> [A] man ought to do something if it is his duty to do that thing. Equally, he ought to do it if he is under an obligation to do it, and he ought to do it if it is right, in view of some moral principle to which he subscribes, that he should do it.[6]

Lemmon tells us that the relevant duties include those incurred in virtue of one's job; e.g., duties as policeman, as headmaster, or as garbage collector.[7] We are "under an obligation" in the relevant sense when we make a specific commitment, e.g., by signing an agreement.[8]

Moral obligations are not that easily incurred, however. An extreme example makes the point most vividly. Consider the exe-

cutioner in some horrendous death camp. He has duties in virtue
of being executioner. But they are not moral duties; they are merely
part of the job. Nor do they become moral obligations if he has committed himself to killing his victims, nor if he subscribes to the
moral necessity of his work.[9] So we should not accept moral dilemmas on the basis of these principles that Lemmon offers. They are
false.

Williams, Marcus, and van Fraassen all find support for moral
dilemmas in cases where moral sentiments play a prominent role.
Williams' defense depends upon the legitimacy of regret in some
cases where the agent has acted for the best. He appeals to the example of Agamemnon at Aulis. Agamemnon is told by a seer that he
must sacrifice his daughter to satisfy a goddess who is delaying at
Aulis his expedition against Troy. Agamemnon accepts that his
duty to carry out his mission takes priority over sparing his daughter, and he sacrifices her. As they bear on his example, Williams'
observations tell us that Agamemnon may well have regretted his
terrible deed in part because of his belief that he ought not to have
done it, even though it was morally best.[10] Williams points out that
it can be admirable, and far from irrational, to feel such regret.[11]
And he maintains that it does not account for those things to hold
that the duty to spare the daughter was merely *prima facie*. His reason for this last contention is that a *prima facie* duty makes a claim
for consideration as "the only thing that matters" in the case,

> . . . and if a course of action has failed to make good this claim in a
> situation of conflict, how can it maintain in that situation some resid-
> ual influence on . . . moral thought?[12]

This objection to appealing to the notion of *prima facie* duty
clearly fails. A *prima facie* duty does not make a claim to be the
only thing that matters in a situation. Its claim is to be something
that matters. What yields a *prima facie* duty is, in W. D. Ross's
words, "a definite circumstance which cannot seriously be held to
be without moral significance."[13] So such an act can have moral
importance when something else matters more and is obligatory.
Thus, foregoing a *prima facie* duty *can* be a source of reasonable
regret.

There is another dilemma-free way in which we can account for
the sort of regret that Williams discusses. We can suppose that Agamemnon subscribed to a moral code that he had no reason to question, on which both sacrificing and sparing his daughter were oblig-

atory. That would be a flaw in the code, but nothing that would place it beyond reasonable belief. If we like, we can say that we thereby acknowledge moral dilemmas as "facts of rational moral psychology." That gives very little to advocates of the real possibility of moral dilemmas. What they need is reason to think that some rationally acceptable dilemma-prone moral code might possibly be true. In analogous terms, the best of the semantic paradoxes show that inconsistent beliefs are "facts of rational doxastic psychology." Reasonable thinkers have on occasion accepted all the premises of a paradox. That scarcely shows that an inconsistency is possibly true. Returning to the moral case, if Agamemnon's regret were partly prompted by a reasonably believed failure of obligation, it would have been rational and admirable. No actual dilemma is required.

There is another salient ground for regret in Agamemnon's case—the dreadful fact of his having killed his innocent, beloved daughter. No need for him even to *think* that he was obligated *not* to do that. It is regrettable enough that he *was* obligated to do something that bad, and that he could do nothing better. In fact, the story does not have Agamemnon recognize any obligation not to sacrifice his daughter. It does have him lament, "Which of these courses is without evil?"[14] as he weighs his alternatives. The basis for that lament is sufficient justification for regret.

Williams claims that a dilemma-free morality "eliminates from the scene the 'ought' not acted on."[15] That is true on the reading where "ought" expresses absolute moral obligation. When such an "ought" is obeyed, no alternative one is neglected. But Williams is wrong in claiming that dilemma-free moralities must have it that obedience in cases like Agamemnon's occasions only feelings of "relief (at escaping mistake), or self-congratulation (for having got the right answer), or possibly self-criticism (for having so nearly been misled)."[16] One who believes in a dilemma-free morality can reasonably feel regret in cases where adhering to it had harmful results. I conclude that Williams' assertions about regret offer no substantial support for moral dilemmas.

Marcus and van Fraassen acknowledge that dilemma-free moralities permit justification for regret.[17] But they contend that some instances where the moral sentiment of guilt is appropriate do supply evidence for moral dilemmas. Marcus holds that even when the act chosen is clearly superior, it is

... inadequate to insist that the feelings of guilt about the rejected alternatives are mistaken, and that the assumption of guilt is inappropriate.[18]

Marcus claims that those who deny moral dilemmas do hold that such guilt is mistaken and inappropriate, and that she calls "false to the facts."[19] Making a similar point, van Fraassen asserts, "it is appropriate to feel guilty if and only if one is guilty."[20] Thus we are to infer that those who feel appropriate guilt while having acted for the moral best must *be* guilty of having violated another moral duty.

What goes into this emotion of moral guilt? It at least includes believing that one has somehow failed morally, and feeling sorry about a supposed bad feature of the believed lapse. That belief and feeling are appropriate *to the facts* only if one has not done all that morality requires.[21] But another sort of appropriateness is also relevant here. Feeling guilty is *subjectively* appropriate when the belief that one has failed which prompts the feeling fits one's moral principles. If your convictions include that every debt morally must be repaid, it is appropriate to your morality for you to feel guilty about defaulting. When someone does what is morally best while neglecting something his morality requires, his feeling guilty is therefore appropriate only because it is called for by morality as he sees it. It does not fit the facts. This sort of appropriate guilt does not imply that a moral mistake has been made. So an opponent of moral dilemmas can consistently hold that feeling guilty about a morally superior act is clearly appropriate at times—but in this subjective way, not in light of any omitted actual moral obligation.

Another sort of case mentioned by both Marcus and van Fraassen is that in which equally strong moral claims demand incompatible actions. Marcus suggests that we consider an example where equally worthy identical twins are in jeopardy, and either, but not both, can be saved.[22] Van Fraassen turns our attention to the character Nora in Ibsen's *A Doll's House.*[23] Near the end of the play, Nora must choose between a duty to her family and a duty to herself that she recognizes to be equally compelling.

We must grant that there are cases where competing moral considerations have exactly the same force. And we must grant that Nora is faced with conflicting duties. So if each of Nora's duties is morally mandatory, or there is an absolute moral obligation to each twin, then the possibility of moral dilemmas has been established.

But there is no need to count each of these alternatives as morally required. We have the familiar option of holding that when moral factors are equal, each act is permitted and none is absolutely obligatory. The sense in which these alternatives are clearly "duties" is not that of being all things considered moral requirements. Not every duty given by one's role in a family is morally mandated. Rather, each is required to carry out a role that has considerable moral significance. Similarly, duties to oneself are requirements for self-fulfillment, which gives them moral importance. We can go so far as to count all such requirements as *prima façie* moral duties. The crucial point is that such requirements have a moral bearing that permits there to be rival conduct having equal moral worth. Thus it is possible consistently to acknowledge that there are conflicts between things properly called "duties," while maintaining that ties for first place between competing moral considerations never generate moral dilemmas.

Van Fraassen considers and rejects this sort of account.[24] He attributes it to the axiologist, that is, one who affirms:

> "It ought to be the case that *A*" is true exactly if some value attaching to some outcome of the alternatives making *A* true is higher than any attaching to any outcome of the alternatives making not-*A* true.[25]

Van Fraassen contends that denying conflicting moral demands in Nora's case amounts to taking it to be

> fundamentally the same as that of the philanthropist who regrets that he has but one fortune to give to mankind and agonizes over the choice between endowing the arts and furthering birth control (or as the case of the revolutionary who agonizes over the question where he should risk his life against oppression—in Bolivia or in Guatemala).[26]

It is important to realize that the axiologist is not alone in denying all moral dilemmas. Ross's avowedly deontological theory excludes dilemmas.[27] And there is no need to appeal to Ross's special technical concept of a *prima facie* duty in order to state a believable morality which is both nonaxiological and dilemma-free. We can do this with permissibility as the only deontic primitive. To illustrate with a simple view, here is a miniscule morality that forbids harming people, and counts promise-keeping and returning borrowed items on time as equal positive qualities of acts:

(M) An act is permissible if, and only if, it harms no one unless harm is unavoidable and it satisfies at least one of the following:
  (i) it is an act of promise-keeping,
  (ii) it is an act of returning a borrowed item on time, or
  (iii) no alternative to it is either an act of promise-keeping or an act of returning a borrowed item on time unless each such act harms someone.

The idea behind a theory of this form is that some features of acts, like being of harm to someone, render an act impermissible when an alternative without such a feature exists. Other qualities of acts, like being a timely return of a loan, render an act obligatory (i.e., not permissibly not done) if the act does not have a forbidding feature and nothing else of the same merit can be done instead. When an equally worthy alternative is available, each of them is permitted. When no choice has a morally significant quality, any choice is permitted. Clearly, (M) is not axiological. It permits and forbids in sheer independence of the values of alternative outcomes. Yet (M) never yields a moral dilemma.

This theory's form poses a noteworthy pedagogical problem. Consider how best to teach a child what is moral. (M) purports to state a necessary and sufficient condition for permissibility. To do that in the manner of (M), we would have to list every morally important quality of acts so that permission could be given to do other things when nothing with such a quality was available. Obviously (M) does not actually do that. But suppose that the truth about moral permissibility is a much lengthier version of (M). Think of how difficult it would be to teach a counterpart of (M) that contains all the relevant conditions. It would be ineffective merely to recommend simple implications of the doctrine like this one: promise-keeping is permitted when no one is thereby harmed. That is easy to convey, but weak. It forbids nothing. It is much more effective to say that promises are not permissibly broken, and loans must be repaid on time. Those simple dicta make the same moral evaluations as (M) except when more than one alternative has a morally significant quality. Even then, if some such maxim is followed, an (M)-like theory permits the act unless it has a forbidding feature as well. Thus even if our best moral insights favor something like (M), it is no wonder that in most moral training such simple (though dilemma-prone) maxims are taught. In any event, we see that con-

sistent theory-based opposition to moral dilemmas is not restricted
to axiologists.

We turn now to the analogies that van Fraassen offers to cast
doubt on the dilemma-free interpretation of Nora's predicament.
He claims that Nora's case is parallel to those of the philanthropist
and the revolutionary only if a feeling of guilt is appropriate exactly
when regret is.[28] It seems that we are to see that guilt is the emotion
appropriate for Nora, while regret is all that fits foregoing one charity or revolution for another. In reply, first it should be repeated that
when we do see things that way we may be discerning appropriateness to certain presumed moral convictions. By assuming their
belief in any of a variety of common moral codes, we can find Nora
appropriately guilty and the other two appropriately regretful. But
again, this subjective appropriateness leaves us free to deny all
actual conflict of absolute moral obligations. Another weakness in
the case for dilemmas based upon these analogies is that they break
down on a misleading contrast. In the example of Nora's predicament (and Marcus' example of the twins in jeopardy), known harm
will come to known appreciated individuals whichever alternative
is taken. Thus sorrow is clearly justified there. In the examples of
the philanthropist and the revolutionary, the choices have unknown
effects, difficult even to guess. Perhaps misgivings fit such evidence
about consequences, but not sorrow. This difference can mislead us
into thinking that the more troubling emotion of guilt fits the facts
in Nora's case (and the twins' case), but not the other two. But this
difference in the strength of warranted discomfort, unlike that
between factual warrant for feelings of guilt and for regret only, does
not imply a difference in moral status. There is no disanalogy here
that supports the possibility of a moral dilemma.

The last claim on behalf of moral dilemmas for us to consider
was made by Marcus. It concerns a salubrious effect of appropriate
ascriptions of guilt:

> [A]n agent in a predicament of conflict . . . will strive to arrange his
> own life and encourage social arrangements that would prevent, to
> the extent that it is possible, future conflicts from arising. To deny the
> appropriateness or correctness of ascriptions of guilt is to weaken the
> impulse to make such arrangements.[29]

Marcus advocates a principle according to which it is invariably
morally obligatory to make such arrangements:

> One ought to act in such a way that, if one ought to do x and one ought to do y, then one can do both x and y.[30]

This principle is inescapably obeyed if there can be no moral dilemmas. But if they can arise, then the principle has it that morality counsels us always to prevent them. So motivation to arrange for their prevention would always incline us toward actions we ought to perform.

It is not inevitably morally best to avoid the sorts of situations where proponents of moral dilemmas find them, however. To take a dramatic example, the spy whose information deters a nuclear conflict may have had to cheat, to betray family and friends, and to abuse innocent strangers in order to get that information. If so, then given that the prevention of nuclear war was at stake, he was morally required to do those things despite the fact that the moral defects of those acts are of sorts that are supposed to create obligations not to perform them. So the proponent of dilemmas would say that the spy faced several of them on that assignment. Yet if the mission was the sole means of averting the disaster and the spy could have arranged to cancel it, then it was morally mandatory that he not do so. Thus not all motivation for preventing circumstances that would be alleged to contain dilemmas works for the best.

Nonetheless, it is usually a good thing to prevent such situations from coming about. And being troubled about a believed moral failing does motivate such prevention. But there is no need to attribute moral dilemmas in order to justify powerful negative sentiments in these cases. In our examples of Nora and Agamemnon, evils ensue whatever is done. To say that guilt is not in fact appropriate is not to imply that only weaker negative feelings are suitable. Abhorrence of the ensuing evils is fitting. And that emotion equally inclines us to prevent circumstances from arising in which something so harmful must be done. So without presupposing any conflict among absolute obligations we can still ascribe and justify feelings that help to keep us out of these predicaments.

We have seen no instance in which attributing a moral dilemma improves our understanding of the case. Moral dilemmas are of no special assistance in accounting for moral sentiments or in promoting good behavior. And as we saw at the outset, their existence would confound us with the prospect of impermissible obligations. The reasonable conclusion is that they are impossible.[31]

Notes

1. E.J. Lemmon, "Moral Dilemmas," *Philosophical Review,* 71 (1962), 148 [p. 105 in this volume].

2. Ruth Marcus, "Moral Dilemmas and Consistency," *Journal of Philosophy,* 77 (1980), 126 [p. 193 in this volume].

3. Bernard Williams, "Ethical Consistency," *Problems of the Self* (Cambridge: Cambridge University Press, 1973), pp. 173–76 [pp. 122–26 in this volume].

4. Bas van Fraassen, "Values and the Heart's Command," *Journal of Philosophy,* 70 (1973), 15 [pp. 148–49 in this volume].

5. Van Fraassen, p. 12 [pp. 145–46 in this volume].

6. Lemmon, p. 140 [p. 102 in this volume].

7. Lemmon, p. 140 [p. 103 in this volume].

8. Lemmon, p. 141 [p. 103 in this volume].

9. Indeed, we can use this kind of case to see that the feature of being required in one's line of work has no moral force at all by itself.

10. Williams, p. 174 [pp. 123–24 in this volume].

11. Williams, p. 173 [p. 123 in this volume].

12. Williams, p. 176 [p. 126 in this volume].

13. W.D. Ross, "What Makes Right Acts Right?" *The Right and the Good* (Oxford: Clarendon Press, 1930), p. 20 [p. 87 in this volume].

14. Aeschylus, *Agamemnon,* tr. Eduard Fraenkel (Oxford: Clarendon Press, 1950), line 211.

15. Williams, p. 175 [p. 125 in this volume].

16. Williams, pp. 175–76 [p. 125 in this volume].

17. Marcus, pp. 131–33 [pp. 197–99 in this volume]; van Fraassen, p. 14 [pp. 147–48 in this volume].

18. Marcus, p. 131 [p. 197 in this volume].

19. Marcus, p. 130 [p. 197 in this volume].

20. Van Fraassen, p. 13 [p. 147 in this volume].

21. It has been suggested that omitting a *prima facie* duty may be sufficient to make a feeling of guilt objectively appropriate. I am inclined to believe rather that feeling guilty fits the facts only if one really is guilty of wrongdoing, while regret is the appropriate emotive response to failing to carry out what is just a *prima facie* duty. But if the suggestion is correct, then we have another way in which guilt feelings can be appropriate while no absolute moral duty is left undone.

22. Marcus, p. 125 [p. 192 in this volume].

23. Van Fraassen, p. 9 [p. 142 in this volume].

24. Van Fraassen, p. 14 [pp. 147–48 in this volume].

25. Van Fraassen, p. 7 [p. 140 in this volume].

26. Van Fraassen, p. 14 [p. 148 in this volume].

27. Ross counts each most stringent *prima facie* duty as something that it is morally right to fulfill (Ross, p. 41 [pp. 98–99 in this volume]).

28. Van Fraassen, p. 14 [p. 148 in this volume].

29. Marcus, p. 133 [p. 199 in this volume].

30. Marcus, p. 135 [p. 200 in this volume].

31. I am grateful for comments on a previous version of this paper from Eva Bodanszky, Herbert Heidelberger, and the editors and a referee for the *Philosophical Review*.

13

Moral Realism and Moral Dilemma

Philippa Foot

Two articles written by Bernard Williams in the middle sixties have recently been receiving attention in the journals, and as these articles argue against moral realism the idea is abroad that moral realism is under attack. Moreover, the basis of the attack seems to be new. Formerly, emotivists and prescriptivists started out from pervasive features of moral language to draw a contrast between moral judgments and descriptions of the world; but the latter day antirealism bases itself rather on a set of special cases: those in which moral judgments seem to collide. The suggestion is that moral realism or cognitivism (which are not here distinguished) cannot do justice to the facts of moral dilemma or conflict. So in "Ethical Consistency"[1] Williams insisted that the feelings we have in situations of moral conflict show that the "structure" of moral judgments is unlike that of assertions expressing beliefs. In "Consistency and Realism"[2] he argued that we can tolerate inconsistency in moral principles though not in assertions, and that this is explained by the

From *The Journal of Philosophy* 80 (1983): 379–98. Reprinted with the permission of *The Journal of Philosophy* and the author.

fact that it is the concern of the latter but not of the former to reflect an "independent order of things."

I shall suggest that in each of his papers Williams' main line of argument is mistaken. Much of what he says in "Ethical Consistency" about moral conflicts not being resolvable "without remainder" seems to me to be true, but in no way inimical to moral realism. Much of what he says in "Consistency and Realism" about the tolerability of conflicting moral principles also seems to be true, but again the antirealist or anticognitivist inference seems mistaken. This is what I shall try to show.

I

To open the discussion we must, of course, ask what it is that Williams, and others who have recently written about moral conflict or moral dilemma, are talking about. We notice at once that cases of "moral dilemma" are not necessarily ones in which anyone is *in* a dilemma about what to do; and that the conflict is not the war that goes on in someone's mind when he is torn between alternatives. For while in many favorite examples the protagonist is torn; like Agamemnon who must sacrifice his daughter to save his campaign, or Sartre's youth caught between his duty to his mother and to the cause of freedom, it is unlikely that anyone who has to break a promise to see a friend, in order that he may save a life, should be in conflict about what to do. Yet the latter is also treated in this discussion as an example of moral conflict. The conflict in question is therefore between principles such as *keep promises* and *save lives,* not the conflict that may be produced in someone's mind by these clashes of principle.

In situations of moral conflict as thus understood one principle enjoins one action and another another, and it is impossible that the agent should do both. Usually the impossibility is fairly and squarely an empirical impossibility, and it is often up to the agent to rack his brains for a way out before declaring that the conflict is real. But sometimes the mere descriptions of the obligations are such as to rule out joint compliance, as when one has promised to say nothing and also to tell all. In one way or another joint compliance is ruled out, and this is what matters in the present context. Williams prefers to describe conflict situations in terms of a and b and the impossibility of doing both a and b. For the sake of con-

venience I shall, however, normally talk about the fact that someone ought to do *a* and ought to do *not a*. I shall treat 'X ought not
to do *a*' as a mere stylistic variation on 'X ought to do *not a*'. With
'It is not the case that X ought to do *a*', which is of course quite
different, I shall not be directly concerned.

As already mentioned Williams argues in "Ethical Consistency" that cognitivist accounts of moral judgment cannot do justice to the facts of moral conflict, especially as these facts concern
the feelings attending choice in conflict situations. He writes:

> It seems to me a fundamental criticism of many ethical theories that
> their accounts of moral conflict and its resolution do not do justice to
> the facts of regret and related considerations: basically because they
> eliminate from the scene the *ought* that is not acted upon. A structure
> appropriate to conflicts of belief is projected on to the moral case; one
> by which the conflict is basically adventitious, and a resolution of it
> disembarrasses one of a mistaken view which for a while confused
> the situation. Such an approach must be inherent in purely cognitive
> accounts of the matter; since it is just a question of which of two con-
> flicting *ought* statements is true, and they cannot both be true, to
> decide correctly for one of them must be to be rid of error with respect
> to the other—an occasion, if for any feelings, then for such feelings as
> relief (at escaping mistake), self-congratulation (for having got the
> right answer), or possibly self-criticism (for having so nearly been
> misled).[3]

Williams says that moral conflicts are not all soluble without
remainder and that this is implied by the regret that we sometimes
feel in conflict situations even when we are convinced that we acted
for the best. Two principles compete with each other for the guidance of our conduct and the one whose claim is rejected is not simply eliminated like a belief that we decide is false. The reality of the
claim that we judge less pressing goes on making itself felt in the
form of regret, and in the form of other phenomena such as our
willingness to make it up to anyone who is injured by our choice.
In this, he thinks, moral conflicts are more like conflicts of desire
than of belief. If I decide that one of two conflicting beliefs is true
the other cannot substantially survive the decision, because to
decide that a belief is untrue is to abandon that belief. A rejected
desire, on the other hand, may survive the decision not to satisfy it,
and even when opportunity for satisfaction is past it may reappear
in the form of a regret for what was missed.

Thus Williams believes that one cannot accept a picture of

moral judgment which as he puts it "makes it a necessary consequence of conflict that one *ought* must be totally rejected in the
sense that one becomes convinced that it did not actually apply."[4]
And this account he had said, in the passage quoted earlier, to be a
necessary consequence of cognitivist theories of ethics. I shall first
consider the argument from feelings to "remainder," then the facts
about what is left over after the solution of a moral conflict, and
finally the implications, if there are any implications, for the debate
about moral realism.

The feelings that Williams thinks so significant are the feelings
of regret or "distress" that someone may feel even when he is convinced that his choice of actions was morally justified—that he was
"acting for the best." The regret or distress is supposed to show that
the agent thinks he was doing something he *ought not* to have done,
e.g., in breaking a promise, even where he also thinks that in the
circumstances it was right to break the promise. Opting in favor of
the second *ought* leaves the other intact. Against those who would
call the feelings irrational Williams says that they do the agent credit
by showing that he takes promises seriously. He also insists that the
regrets spring from the agent's thought that he has done something
that he ought not to have done and not from some other distressing
feature of the situation.

The form of this argument is surely strange. It is not, of course,
to be denied that feelings such as regret are "propositional," so that
to feel regret is at least to feel *as if* something in some way *bad* has
happened. But it does not follow that it has happened, and perhaps
not even that the subject thinks it has, since one may say, e.g., "I
can't get away from a feeling of regret, though I know I haven't
actually lost anything at all." It is impossible to move from the existence of the feeling to the truth of the proposition conceptually connected with it, or even to the subject's acceptance of the proposition.
Nor will it help to argue, as Williams does, that if we call such feelings irrational we must be using the description nonpejoratively. He
thinks it creditable for someone to feel "moral distress" when making a choice in a conflict situation; to which one would reply that
this is neither here nor there. There are plenty of feelings which are
irrational without being discreditable, as for instance feelings of
guilt about giving away the possessions of someone lately dead. It
would obviously be wrong to conclude from the fact—the normality
almost—of such feelings that there was indeed some element of
wrongdoing involved.

It is a mistake then to think that the existence of feelings of regret could show anything about a remainder in cases of moral conflict. The feelings are rational feelings only if it is reasonable to think that given a conflict situation there *is* something regrettable or distressing even in a choice that is clearly right. What we find is, I think, that there may indeed be a "remainder" in the shape of obligations unfulfilled, and things left undone which it is correct to say that we ought to have done. But whether it is always *regrettable* or *distressing* when obligations are unfulfilled or things left undone which ought to be done is more doubtful. I shall come back to this point after discussing the obligations and *oughts* which are left over when a moral conflict is resolved.

Suppose a case in which I have promised to do an action *a* and promised to do an action *b;* say to be best man at A's wedding and also at B's. By bad luck A and B fix their weddings for the same day and I can't attend both. So one of my promises must be broken, and as we are considering resolvable conflicts we may suppose that for some reason my promise to A has clear precedence over my promise to B. Nevertheless I promised B, and nothing has happened to release me from this promise. I have an obligation to him which, when I opt for A's wedding, I decide not to fulfill. I still have the obligation and it doesn't much matter at what point it will be right to say that I *had* it rather than that I *have* it. In one form or another the obligation stands, unless B releases me from it before the time for fulfillment is past.

There is, therefore, the possibility of saying truly that I have an obligation to do two things that cannot both be done. I have an obligation to do each of them, although I do not of course have an obligation to do them both.[5] It can similarly be true that I *ought* to do them, though it is less common to speak of two things each of which I *ought* to do but both of which I cannot do, than of two mutually exclusive obligations. However, the possibility of employing 'ought' like this reminds us that the area of conflict-without-inconsistency extends beyond that part of morality which has to do with what are strictly obligations. Moreover we find the phenomenon outside morality, since it makes perfectly good sense to say, when pressing business has given one overriding reason to go to town, that one nevertheless *ought* to be at home nursing one's awful cold. Indeed it may seem surprising that anyone should ever have denied that I can have an obligation to do *a* and an obligation not to do *a,* or that I ought to do *a* and ought not to do it. Why is this harder to accept than the fact that I can have two engagements that conflict? 'Engagement' like 'obligation' and 'ought' is an "action-guiding" term: if people did not look in their books or consult their memories and say things like "I have an engagement *so I must go . . .* " the concept of an engagement would not be known. But I can have conflicting engagements as I can make conflicting promises. Given the possibility of lying promises and of memory lapses I may even have promised that I will do *a* and that I will not. If I cannot have an engagement to do *a* and not to do *a* this is merely because (i) we do not have negative engagements and (ii) in 'engagement to do *a*' the context of reference of *'a'* is arguably opaque. I can only too easily have an engagement to do *a* and an engagement to do *b,* where I can't do both. Propositions about engagements are what I shall call 'type 1' propositions. We shall see later on that 'obligation' and 'ought' can appear also in another type of statement, but for the moment the comparison with promises and engagements is to be kept in mind.

Why is there resistance to the idea that an obligation or *ought* may be overridden without being destroyed? I think that there are two reasons.

The first reason is that a certain kind of example sticks in one's gullet. Suppose, for instance, that some person has an obligation to support a dependent relative: an aged parent perhaps. Then it may be that he *ought* to take a job to get some money. The obligation produces an *ought* related to means. So far so good. But what if the only means of getting money is by killing someone? The obligation to refrain from taking someone's life—to refrain from interfering with him in this way—is stronger than the obligation to give aid to the parent, and it is therefore clear what should be done. Should we nevertheless say that although the agent ought not to kill he also ought to kill, since *oughts* which are put out of action by stronger *oughts* are not thereby destroyed? Surely this *ought* is destroyed by the superior injunction against taking life?

We need an explanation of why it is not the case that the son or daughter ought to kill to get the money, but an explanation which will not interfere with the general principle that oughts that are inactivated may nevertheless stand. We find such an explanation in the fact that in the problem case the ought would have been attached to the killing only because killing was a *means* to the fulfilling of an obligation. For the only things that count as means are *possible* actions. If some obligation of mine could be fulfilled only by my flying up the ceiling we do *not* therefore say that I ought to fly up to the ceiling. This may seem irrelevant, for we think "After all it is *possible* for the agent to kill and get the money." But then we have failed to notice that it is not only natural law possibility that matters but also moral possibility. When we say "I can't do such and such" we do not necessarily mean that there is nothing in our power that would bring it about. Often we have no idea whether there are some steps we could take if only no holds were barred. In our example killing is not a *possibility* and so there is no question of treating it as a means which *ought* to be taken to fulfill the obligation to support the aged parent. The obligation stands though we cannot discharge it; but it does not make the killing into something that ought to be done. Puzzling as moral modalities are, we have to take them into account if we are going to understand the way concepts such as *obligation* and *ought* work in cases of moral conflict.[6]

The second reason why it is hard to see that conflicting ought statements ('ought *a*' and 'ought ~*a*') are consistent is that there is another kind of ought statement for which this is not true. When 'ought *a*' and 'ought ~*a*' are both of the second kind, are both as I shall say used in type 2 ought statements, they are not consistent. This distinction between types of statements is essential to my argument.

What is a type 2 ought statement? What is it that makes 'ought *a*'(2) inconsistent with 'ought ~*a*'(2), although 'ought *a*'(1) is consistent with 'ought ~*a*'(1)?

The explanation is that type 2 ought statements tell us *the right* thing to do, and that this means the thing that is *best* morally speaking, or speaking from whatever other point of view may be in question. It is implied that for one for whom moral considerations are reasons to act there are better moral reasons for doing this action than for doing any other. As this cannot be true both of *a* and of ~*a*, 'ought *a*'(2) is inconsistent with 'ought ~*a*'(2). 'Ought *a*'(2) is not, however, inconsistent with 'ought ~*a*'(1). I can have reason not to do something and yet have better reason to do it than I have to do anything else.

From the fact that 'ought *a*'(1) is consistent with 'ought ~*a*'(1) but that 'ought *a*'(2) is not consistent with 'ought ~*a*'(2) it follows that if we are to keep the intuitively sound idea that what one ought to do one is permitted to do it is in terms of type 2 ought statements and not those of type 1 that permissibility must be defined. So '*a* is permissible' = def. 'It is not the case that *a* ought not to be done' (type 2 ought). And 'ought *a*'(2) does imply ~(ought ~*a*)(2).

The division into type 1 and type 2 propositions, as I am describing it, belongs also to some other action-guiding statements. The test of whether an action-guiding predicate appears, or is here appearing, in statements of one type or the other is that it does or does not make sense to add $\emptyset(\sim a)$ to $\emptyset a$. By this test we would determine that 'dangerous' on its own appears only in type 1 statements, since the assertion that it is dangerous to do a is always consistent with the assertion that it is dangerous not to do a. (It may or may not be *more* dangerous to do a.) If it is dangerous to pick up a poisonous snake it may nevertheless be dangerous not to pick it up if that is the only way of getting rid of it. By contrast 'imprudent' is a predicate appearing only in type 2 propositions: if it is imprudent to do a it cannot be imprudent not to do it.

From this last example we might conclude that a type 2 statement is one asserted on an 'all things considered' basis, while a type 1 statement is not. But this would be wrong. 'All things considered it is dangerous' contrasts with '*Prima facie* it is dangerous' and *both* are about a dangerousness of doing something which is compatible with the dangerousness of not doing it. And type 1 obligation statements can also be said either to be *prima facie* true or to be true all things considered. For instance there is a *prima facie* case for thinking that I have an obligation to support someone in old age given that he is my father. But perhaps he deserted my mother before I was born, and then I may say that all things considered I have no obligation to him. But even if, in rather different circumstances, I think that all things considered I do have an obligation to this man it does not follow that the right thing to do would be to support him; because my obligation to my children may override my obligation to him. So it was only a type 1 obligation statement that was said to be, all things considered, true.[7] Yet we do naturally indicate that we are employing 'ought' in a type 2 statement by talking about what ought, all things considered, to be done. Why is this? I think it is simply because the same word 'ought' is used in both kinds of statements, and type 2 propositions take as evidence all the available type 1 statements about the same subject together with any principles for deciding priorities. Therefore, in moving from the consideration of any type 1 ought proposition to a type 2 ought proposition about the same subject we can mark the transition by asking "And what all things considered ought we to do?"

We now have a clue to the relationship between type 1 and type 2 statements about what ought to be done. A type 1 proposition says what *is* the case not what *prima facie* is the case. But a reference to what it states to be the case can *appear in* a *prima facie* statement about a type 2 predication. So the fact that something is dangerous shows that *prima facie* it is imprudent to do it. And the fact that I have an obligation (1) to do something shows that *prima facie* I ought (2) to do it.

We have now seen one definite way in which even the clearest solution of a moral conflict, expressed in a type 2 ought statement which nobody doubts, nevertheless can leave "remainder." It may be the case that a "conflicting" type 1 proposition about an obligation, or about what the agent ought to do, or to have done, is true. Let us now return, briefly, to the question of feelings such as regret or "distress." Does the truth of a type 1 obligation or ought statement make such feelings rational? This, as I said before, does not seem to me as clear as Williams and others suppose.

Take, for instance, the breaking of a promise. One has promised to meet someone but must instead take an accident victim to hospital. Obviously one regrets it if there isn't time to let the promisee know and he is incommoded. But this is regret for a consequence, not regret for the breaking of a promise as such. So, to isolate the latter we shall suppose that things turn out splendidly all around; the promisee does not have a moment's annoyance, and meets his future beloved, or someone who offers him a job, while standing at my door. Are we to say that nevertheless in the general rejoicing there should be an element of distress (moral distress) because after all a promise was broken and that is something bad, and therefore regrettable? To this suggestion one hardly knows what to reply. A charitable man would wish for a world in which promises did not have to be broken (as he would wish for a world in which there were no earthquakes) but this is because the breaking of promises, even when necessary, usually has bad effects. And we may perhaps suppose that a moral man must regret the wickedness of the world that contains wantonly broken promises and must regret it even when nothing bad follows. But our case is one in which neither bad consequences nor wickedness is there to be regretted. Someone who nevertheless felt distress would seem to many of us rather foolish, though one can of course imagine a society in which someone was held to incur *shame* or be "tainted" by having to break a promise even for the best of moral reasons. And we ourselves might feel a distress that we thought rational if for instance we had had to reveal a secret that we had promised we would never tell, even if no harm had come of it or could come of it. If the secret were something that its possessor had wanted very much to guard then even if his reputation had been enhanced rather than damaged by the telling of it, and even if he were now dead, one might still feel that it was something regrettable that one had to do, and one might hate to do it. The area seems to be one of uncertainty, but perhaps this very fact makes the argument from feelings to "remainder" in the solution of moral conflicts a bad one. And it is in any case unnecessary since the language clearly allows us to use 'obligation' and 'ought' as we use 'engagement' in type 1 statements.

Against one argument from feelings one must definitely protest. This is an argument, not given by Williams himself, about feelings of guilt. It may seem obvious that no one can be guilty if he acts "for the best," but this has recently been denied by Bas van Fraassen and by Ruth Marcus, though on slightly different grounds. Van Fraassen questions the connection between guilt and what is in the agent's power, suggesting that it is denied in the doctrine of original sin.[8] Ruth Marcus takes a line which seems to me more interesting.[9] She says that while no one is guilty except by virtue of a choice he has made, this does not rule out the possibility that he is in such a situation that he will be guilty if he does some action and guilty if he does not, since he is free to choose to do it and free to choose not to do it.

These arguments seem to me wrong, but they throw upon their critics the burden of showing why guilt is not like this. After all shame is not, since no one has to be in any way responsible for what brings shame on him, which might in Dostoevskian fashion be his craven or dissolute father. And then there is the idea of "dirty hands," which is perhaps nearer to what Marcus wants, since one's hands get dirty only by one's voluntary actions—or so one may suppose—but nevertheless the situation may be such that no one can emerge with clean hands whatever he does. Perhaps he must either betray his friend's confidence or let an innocent man be condemned through his silence. Either action seems shabby and what Williams has called the moral "disagreeableness" will not go away, even if there is a clear solution and the agent is guided by it.[10]

What is it then about guilt that makes this different? Since the argument is not about a word I don't want to dispute about the exact boundaries of the concept of guilt. It will be enough to establish that there is some notion of *fault* which is such that no one can be at fault both if he does *a* and if he doesn't do *a*, unless of course the fault is imputed to him on account of some prior choice, or other past or present moral failing.[11]

That there is such a notion of fault is shown by the fact that there is an imputation against which not only physical or mental but also moral necessity is a shield. How can this be denied? There is a clear place for the plea 'I couldn't help it' (couldn't help breaking the promise because I had to attend to the accident victim, and so on). Nor does this plea simply plead mitigation, as if the offense of breaking the promise was merely lessened. If you suffer because I cannot get to the appointment I have with you, I say that I am sorry, meaning that I regret it; but if it was not my fault I do not apologize, and I certainly do not have to "make restitution" as some have suggested. If I *can't* keep the appointment it isn't my fault that you suffer, and it doesn't make any difference whether the necessity of breaking the promise was physical, mental, or moral.

I do not know quite what Marcus, for instance, would say here. Would she deny the whole procedure of refusing to admit moral fault on grounds of moral necessity? Or would she say that it *sometimes won't be allowed,* as if the plea will only do for cases in which no grave moral issues are involved? It is hard to see why this would be more plausible than so restricting the plea of physical or psychological necessity. The plea of moral necessity is that one had to do *a* because it would have been worse not to. Since we are still dealing with cases in which moral conflicts are resolvable, i.e., where it can be said that *a* is worse than ∼*a* or vice versa, and since this can be said no less in grave situations than in others, either *a* is worse than ∼*a* or the other way around. And there seems no basis for driving a wedge between this and moral necessity, or between moral necessity and the absence of fault. This is not of course to deny the suggestion that there are some actions of which it cannot ever be said that I had to do them, pleading moral necessity. But this is only to say that some things never can be such that it is morally worse not to do them than to do them and my argument is neutral about this possibility. What I am arguing is that *if* ∼*a* is worse than *a,* then there is a kind of fault that cannot be imputed to an agent who does *a.* Nor does it seem reasonable to deny this kind of faultiness the name of guilt.

Let me now sum up the results of this first part of the discussion. We asked ourselves a question about the solution of moral conflicts, namely whether they could be solved only "with remainder." What emerged most clearly were some facts about obligations and *oughts:* that an obligation is not annulled by being overridden, and that it is possible to say that a subject ought to do something, even when a more pressing claim makes it impossible for him to do that thing. I tried to sort out various consistencies and inconsistencies which depended on the *type* of ought proposition intended in the words 'X ought to do *a*'. It emerged that 'I ought to do *a*' is in many cases, though not always, consistent with 'I ought not to do *a*'.

How is all this relevant to theories of ethics, to cognitivism and noncognitivism or realism and antirealism? How do things stand with Williams' arguments from the facts about moral conflict to noncognitivism? The first point to notice is a crucial difference of terminology between Williams' articles and the foregoing discussion. On the whole he talks about moral *conflict,* which is how he describes a clash between moral principles where the clash is brought about simply by the facts. But he also talks about *inconsistency* in cases where the clash is between logically incompatible alternatives, either *a* and *not a,* or *a* and *b* where the descriptions of *a* and *b* are such that in no possible world are *a* and *b* both performed. (He has a parallel vocabulary for describing desires which cannot both be satisfied: one has either *conflicting* desires or *inconsistent* desires.) I find this piece of terminology unhelpful for reasons that will become obvious as I go on. In my vocabulary propositions, moral or otherwise, are inconsistent only if being either contradictories or contraries they "contradict" each other.

In the first of his two articles, "Ethical Consistency," Williams' procedure is as follows. He first contrasts conflicts of desire with conflicts of belief and then argues that moral conflicts share characteristics with the former rather than the latter, which is what he thinks his cognitivist opponent cannot admit. Moral conflicts are like conflicts of desire in that they are not settled "without remainder." The decision to satisfy the first of two desires when both cannot be satisfied does not in itself extinguish the second, which may linger in the form of a desire, or of a regret for what was sacrificed. And similarly, the decision to let one "ought" or "obligation" proposition guide one's conduct may leave a remainder whose presence is, he says, indicated by moral regret or distress. In the case of beliefs, however, the decision that one of two conflicting beliefs is true means that the other has been abandoned. So, he thinks, the "structure" of moral judgments is unlike that of the assertions by which beliefs are expressed. And so moral cognitivism is false.

What shall we say about this argument? First of all, that the structure of moral judgments about what ought to be done is like that of statements about what is desirable, and like expressions of desire, insofar as all allow of the 'because of this $\emptyset x$ but because of that $\emptyset(\sim x)$' form of proposition, and therefore of '$\emptyset x$ and $\emptyset(\sim x)$'. But this has nothing to do with cognitivism and noncognitivism. If statements about what is desirable or expressions of desire cannot be understood as being "about the world" this is not why they cannot be so interpreted. It was observed earlier that we may have an engagement to do a and an engagement to do b where it is not possible to do both a and b, and that the decision, however well justified, to keep one rather than the other does not "destroy" the other, which stands as "remainder." We should not, however, be impressed by an argument against a cognitive theory of engagements based on the dissimilarity in the matter of remainders between "conflicts of belief" and "conflicts of engagements." It is usually a plain matter of provable fact that I do or do not have an engagement; if I say that I do have one my assertion is fully licensed by the evidence and however it may be with moral judgments, statements about engagements undoubtedly express beliefs. They differ from certain other cognitive affirmations as judgments about what is dangerous differ from judgments about what is imprudent. They are action-guiding propositions of a type 1 variety. But this, as the case of engagements shows, is a difference found within the class of assertions (assertions about the world) and has nothing to do with the demarcation of that class.

Williams thought that the cognitivist must believe that when two ought statements conflict one is necessarily false, and would therefore find himself unable to explain the fact of regret for an unfulfilled obligation such as a promise not kept. It does not seem to have occurred to Williams that his opponent could simply *allow* the truth of 'I ought to do a' and 'I ought to do $\sim a$' and other consistent propositions whose consistency is easily explicable on a "because of this . . . , but because of that . . . " basis. And the reason it did not occur to him was, it seems, that he thought the cognitivist was committed to a comparison between moral conflicts and conflicts of belief. But why should the cognitivist ever accept such a comparison?[12] Beliefs that conflict are beliefs that *contradict* each other, either directly or in the context of other beliefs. But the whole point about statements about what is desirable and what ought to be done is (for both) that there is a class of statements which *conflict*, in that they give conflicting guidance for action, but which nevertheless can *both be true*. The strange thing about what Williams wrote in "Ethical Consistency" is that a great deal of it seems designed to show exactly this: that moral conflict does not imply "contradiction." It is as if he himself showed the cognitivist how to avoid the very error he thinks the cognitivist must make.

I conclude that it is not "inherent in purely cognitivist accounts" of moral judgment to think that only one of the conflicting ought statements can be true in a case of moral conflict. The cognitivist can do justice to facts about "remainders" and about "moral regret" or "moral distress" exactly as well as anyone else.

II

So far we have been considering examples of moral conflicts which are *resolvable,* though not "without remainder." In such situations there is an answer to the type 2 question 'What, all things considered, ought NM to do?' although this answer may leave standing any number of type 1 ought statements, some of which may *conflict* with the first answer in that both cannot be the basis for action. I have argued that Williams and others are right to insist on this type of "remainder" but wrong to think that an argument has been provided against moral cognitivism or realism.

This has, I believe, disposed of Williams' case against moral cognitivism as it was presented in "Ethical Consistency." But we have yet to look at the points made in the second article, "Consistency and Realism" which seem to be different.

In "Consistency and Realism" Williams first contrasts demands for "consistency" of beliefs and demands for "consistency" of imperatives, suggesting that beliefs claim to represent reality and therefore cannot without error be inconsistent. It cannot be the case that two inconsistent beliefs are both equally good. This could however be so with inconsistent imperatives. There are, at most, pragmatic, practical, reasons for avoiding issuing inconsistent imperatives or other directives; since imperatives are not about the world. In the final paragraphs of the article he argues that to compare moral judgments with imperatives in this respect is to give sense to an antirealist theory of ethics.

> For a realist view would seem to determine a different view of consistency in ethics from that determined by a nonrealist view. . . . on a realist view, the significance of consistency, and the explanation of it as an aim, are going to come down to the simple point that moral

> judgments being straightforwardly assertions, two inconsistent moral
> judgments cannot both be true, and hence (truth being the aim of
> assertions) cannot both be acceptable: one of them must be rejected;
> its reasons must be defective; something must be wrong. . . . But . . .
> the nonrealist approach may well allow for the possibility that one
> can be forced to two inconsistent moral judgments about the same
> situation, each of them backed by the best possible reasons, and each
> of them firmly demanding acceptance. . . . [13]

He gives no argument for thinking that this is how it is with
moral judgments, simply referring back to "Ethical Consistency"
for his reasons for favoring a nonrealist theory. As we have found
these to be bad reasons the comparison between moral judgments
and imperatives must be considered on its merits. The essence of
Williams' case is, clearly, that there may be two equally acceptable
but inconsistent moral judgments just as there may be two equally
acceptable but inconsistent imperatives; there may be nothing
wrong with either of the moral judgments just as there may be nothing wrong with either imperative, and he goes on to say that if we
do favor moral systems in which such inconsistent judgments do
not appear this is not because moral judgments are straightforward
assertions which claim to mirror reality, but rather because we have
pragmatic reasons for doing so. We are to compare moral judgments
to imperatives in order to be able if we so wish to accept "inconsistent moral judgments about the same situation."

To understand this argument we must once more ask what Williams means by 'inconsistent' in the expression "inconsistent moral
judgments"; and when we do so we find that the interpretation of
inconsistency is just what the argument is about. There are two possibilities. The first is that 'It is wrong to do a' and 'It is wrong not
to do a' are inconsistent in the way that straightforward assertions
are inconsistent. The second is that the propositions are inconsistent only as 'Do a' and 'Do not do a' are "inconsistent," that is in
their practical aspect, as enjoining inconsistent states of affairs. On
the first interpretation inconsistent moral judgments contain a contradiction, and cannot both be asserted because reality cannot be
such as to accommodate them both. On the second interpretation
the "inconsistent" moral judgments merely conflict as "inconsistent" orders conflict; and just as such orders *may* be issued so may
conflicting moral judgments be voiced. Although Williams himself
does not put it this way we may see him as arguing that there are
moral situations which by forcing inconsistent moral judgments

upon us make it necessary for us to interpret this inconsistency as conflict rather than contradiction.

What should we think about this argument? The first thing to notice is that it is an argument that cannot be dealt with in the old way, i.e., by showing that where Williams sees inconsistent moral judgments we have in truth consistent assertion of type 1 obligations or *oughts*. For Williams is not here arguing from the remainder that exists even where a moral conflict is most clearly resolvable—where it is most obvious that morally it would be better to do *a* than not to do *a,* or the other way round. He is now arguing from special cases in which no such resolution is in sight. The presence of this different argument is marked by insistence on the difficulty of resolving hard cases such as that of Agamemnon's cruel dilemma, an insistence that would be irrelevant if "remainder" were all that was in question. It seems clear that this is how we should interpret the passage in "Politics and Moral Character" in which Williams refers to "cases of tragic choice where one might say that whatever the agent did was wrong."[14] Moreover, his solution to the problem does allow us some understanding of what it could mean to say that in some situations an agent might be *wrong* whatever he did, as being unjustified in whichever course of action he took, in spite of the fact that a right choice did not exist, and even where fault could not be imputed to him for being in this situation. It is as if the agent made the *wrong choice* although the right choice does not exist. The thought remains puzzling, but the picture can be somewhat filled in, I suppose, by thinking of a subject to whom conflicting orders are issued, and who is penalized by the authorities for disobeying one order or for disobeying the other, whichever it is that he does. Perhaps we sometimes think of ourselves as if we were in this position vis-à-vis "the moral law" and are confirmed in this idea by the discomforts we suffer in certain cases of hard choice.

We see then that Williams' new argument against the moral realist—the one that starts from hard-to-resolve cases—cannot be met in the old way and cannot be dismissed out of hand as unintelligible.

Nor is there any reason to deny the assumption that there sometimes are cases of irresolvable conflict between moral principles or values. Bas van Fraassen writes of irresolvable moral conflict saying "By this I mean a conflict between what ought to be for one reason and what ought to be for another reason, which cannot be resolved in terms of one reason overriding another, or one law or

authority or value being higher than another."[15] And he also refers
to Sartre's contention "that no ethical system can resolve all moral
dilemmas"[16] and to the possibility that there are cases in which "our
morality's guidance is ambiguous, inconsistent, or absent altogether."[17] Marcus writes, in the same vein, "it would appear that
however strong our wills and complete our knowledge we might be
faced with a moral choice in which there are no moral grounds for
favoring doing x over y."[18]

I do not think that there is any reason to deny incommensurability of the kind spoken of by van Fraassen and Marcus, as also
by David Wiggins when he writes "It seems that in the sphere of the
practical we may know for certain that there exist absolutely undecidable questions—e.g., cases where the situation is so appalling or
the choices are so gruesome that nothing could count as the reasonable practical answer."[19] Whether undecidability exists particularly
in situations where the alternatives are ghastly I am not so sure. It
certainly isn't the case that there are *never* better and worse alternatives here, and it is perhaps particularly important to hang on to
this fact, given the temptation to think "The choices are all so awful
morality really doesn't tell me what to do" just because the going is
especially rough. For all I know there may be just as much undecidability in small moral matters, or where the choice is between goods
rather than evils, only it doesn't worry us and we don't notice it so
much.

It is noticeable that although van Fraassen, Marcus, and Wiggins all make references to irresolvable moral conflicts they differ
from Williams in that they do not seem to draw any anticognitivist
conclusion from their observations. The question therefore arises as
to whether Williams has some special assumption which makes him
think such a conclusion irresistible. And whether, if so, his argument can be undermined by challenging this assumption. The
answer to both these questions is affirmative. Williams must be
making the crucial and questionable assumption that in cases of
irresolvable moral dilemma, where the application of one principle
would give the judgment 'there is stronger reason morally speaking
to do a than to do b' and the other '. . . to do b than to do a', and
there is no way of resolving the issue in favor of one rather than the
other, both judgments have to be affirmed. This must be what he
means when he speaks of both as "firmly demanding acceptance."
For as I have suggested the new argument in "Consistency and

Realism"—the one that is based on the existence of irresolvable moral conflict—depends on the idea that contradictory affirmation has to be avoided by a shift from the cognitive to the imperatival mode. And the need arises only if joint affirmation is indeed in question.

What we must ask, therefore, is whether in cases of irresolvable moral conflict we have to back both the judgment in favor of *a* and the judgment in favor of *b*, although doing *b* involves not doing *a*. Is it not possible that we should rather declare that the two are incommensurable, so that we have nothing to say about the overall merits of *a* and *b*, whether because there is nothing that *we* can say or because there is no truth of the matter and therefore nothing to *be* said. The acceptance of incommensurability in morality will of course raise many interesting questions, such as how we recognize it. But incommensurability is not an unfamiliar idea. I think, for instance, of the impossibility of saying in many cases whether one man is happier than another when one lives a quiet and contented life and the other a life that is full of joy and pain. On occasions we make comparisons of happiness with perfect confidence, and yet sometimes there seems nothing to be said. Perhaps we should similarly accept incommensurability in certain cases where conflicting moral judgments suggest themselves to us. And if we do this we do not have to avoid a "contradiction" between two of our affirmations by interpreting these affirmations in a special way.

This concludes my argument against Williams. Let me add two observations by way of postscript. The first is about the interest of the "remainder" thesis. I should say that it is not very interesting in so far as it concerns the fact that obligations that are overridden are not annulled and that there are type 1 propositions about what ought to be done as well as those of type 2. On the other hand the thesis of "remainders" in so far as it is about what is inevitably regrettable in the outcome of moral conflicts is very interesting indeed. The *most* interesting part of the topic has, I think, to do with the inevitable loss involved in a choice between values: when one really good thing which the man of virtue must cherish has to be sacrificed for another, a loss that is often reflected in a conflict of *oughts* or *obligations* but is not described simply by talking about such conflicts. It is Sir Isaiah Berlin who has done most to make us aware of the reality of inevitable loss of one value or another in the political sphere, as when he says, in "Two Concepts of Liberty":

> The extent of a man's, or a people's, liberty to choose to live as they desire must be weighed against the claims of many other values, of which equality, or justice, or happiness, or security, or public order are perhaps the most obvious examples.[20]

David Wiggins too has also recently stressed such things and he has been particularly concerned with the rival goods that a man may see as giving meaning to his life. In neither author does the full difficulty of the most difficult part of the thought about inevitable loss emerge. I mean the thought that so far from forming a unity in the sense that Aristotle and Aquinas believed they did, the virtues actually conflict with each other: which is to say that if someone has one of them he inevitably fails to have some other. Many people do not see the difficulty of this idea because they interpret it rather superficially, as the thought that, e.g., the claims of justice and charity may conflict. But this is easy to accommodate. For insofar as a man's charity is limited only by his justice—say the readiness to help someone by his recognition of this person's right or the right of some other person to noninterference—he is *not* less than perfect in charity. The far more difficult thought is that he can only become good in one way by being bad in another, as if, e.g., he could only rein in his ruthless desires at the cost of a deep malice against himself and the world; or as if a kind of dull rigidity were the price of refusing to do what he himself wanted at whatever cost to others. So Nietzsche found thoughts about the possibility that hatred, envy, covetousness, and the lust to rule must be present in the "general economy of life," and must be "further enhanced if life is further to be enhanced," terrible thoughts; but with his extraordinary and characteristic courage did not decide that therefore they must be false. Perhaps we have shied away from such ideas because we so thoroughly reject Nietzsche. In any case the subject seems a hard one which stands ready to be explored.

I would also add the following comment which has to do with the distinction between cognitivism and realism in ethics. I do not know whether one will be able to distinguish the two unless one understands realism as Michael Dummett does, but if realism is what Dummett meant by realism then it is obvious that they are distinct.[21] A cognitivist theory of ethics says that moral judgments are about the world as are other kinds of assertions—with, no doubt, many significant variations. A realist theory of ethics would be one that refused to let the possession of truth or falsity by a particular ethical proposition stand or fall by our capacity or lack of capacity for assigning truth or falsity to it. Thus the issue of realism, understood like this, has to do with the *implications* of incommensurability. Those who say that where the solution of some ethical conflicts is beyond our capacities there will nevertheless *be* a solution which is perhaps known to God are realists: those who deny it are antirealists as regards the class of propositions in which moral judgments are asserted.

I add this merely as a final comment, because Williams seems to have meant by 'realism' not this but rather cognitivism. Whichever way we understand realism I think his arguments fail.

Notes

This paper was written at the Center for Advanced Study in the Behavioral Sciences at Stanford. I am most grateful to the Center and am also grateful for financial support from the Andrew Mellon Foundation, and for a Fellowship from the National Endowment for the Humanities. Dagfinn Føllesdal, Hans Kamp, Jerrold Katz, Warren Quinn, David Sachs and Richard Watson have all helped me by reading and commenting on various drafts.

1. *Proceedings of the Aristotelian Society,* XXIX (1965) [chapter 6 in this volume].

2. *Proceedings of the Aristotelian Society,* XL (1966). Both articles are reprinted in Bernard Williams, *Problems of the Self* (Cambridge: University Press, 1973) and subsequent references are to this volume.

3. *Op. cit.,* pp. 175–76 [p. 125 in this volume].

4. *Op. cit.,* p. 184 [p. 134 in this volume].

5. See *op. cit.,* pp. 181–83 [pp. 132–34 in this volume] for Williams' denial of what he calls "the agglomeration principle."

6. The relation between obligation and moral possibility is complex. Where an obligation is overridden we may say that it—the obligation—could not be fulfilled. ("I was unable to fulfil my obligation.") But while original obligations may stand it seems that derivative obligations are aborted.

7. In his article "*Prima facie* Obligations," in Z. van Straaten (ed.) *Philosophical Subjects: Essays Presented to P. F. Strawson* (Oxford: Clarendon Press, 1980) John Searle seems to miss the difficulty of identifying what I call type 2 propositions as "all things considered" statements.

8. Bas C. van Fraassen, "Values and the Heart's Command," *Journal of Philosophy,* LXX, 1 (January 1973) [chapter 7 in this volume].

9. Ruth Barcan Marcus, "Moral Dilemmas and Consistency," *Journal of Philosophy,* LXXVII, 3 (March 1980) [chapter 11 in this volume].

10. Bernard Williams, "Politics and Moral Character" in Stuart Hampshire (ed.) *Public and Private Morality* (Cambridge: University Press, 1978) pp. 62–64.

11. A common case would be that in which he had carelessly, or with indifference, *assumed* two obligations likely to conflict. I have throughout ignored this special case.

12. Samuel Guttenplan's otherwise strong article "Moral Realism and Moral Dilemma" in *Proceedings of the Aristotelian Society,* 1979–80, is weakened, I think, by his failure to challenge Williams here.

13. Williams, *Problems of the Self,* pp. 204–05.

14. Hampshire (ed.) *Private and Public Morality,* p. 61.

15. van Fraassen, *op. cit.,* p. 8 [p. 141 in this volume].

16. *Ibid.,* p. 10 [p. 144 in this volume].

17. *Ibid.,* pp. 11–12 [p. 145 in this volume].

18. Marcus, *op. cit.,* p. 125 [p. 192 in this volume].

19. David Wiggins, "Truth, Invention, and the Meaning of Life," *Proceedings of the British Academy,* 1976, p. 371.

20. Isaiah Berlin, *Four Essays on Liberty* (Oxford: University Press, 1980) p. 170.

21. Michael Dummett, *Truth and Other Enigmas* (London: Duckworth, and Cambridge, Mass.: Harvard, 1978) especially the Preface and pp. 1–24 and 145–65.

14

Consistency in Rationalist Moral Systems

Alan Donagan

To the memory of Donald J. Lipkind

In one of the most valuable of his *Essays on Actions and Events,* Donald Davidson[1] has remarked that "life is crowded with examples" of the following sort: "I ought to do it because it will save a life, I ought not because it will be a lie; if I do it, I will break my word to Lavinia, if I don't I will break my word to Lolita; and so on" (34). And of such examples he went on to say this:

> Anyone may find himself in this fix, whether he be upright or temporizing, weak-willed or strong. But then unless we take the line that moral principles cannot conflict in application to a case, we must give up the concept of the nature of practical reason we have so far been assuming. For how can premises, all of which are true (or acceptable), entail a contradiction? It is astonishing that in contemporary moral philosophy this problem has received little attention, and no satisfactory treatment *(loc. cit.).*

Of course, in making these remarks, Davidson had not forgotten the solution of his problem that may be found in the neointui-

From *The Journal of Philosophy* 81 (1984): 291–309. Reprinted with the permission of *The Journal of Philosophy* and the author.

tionist theory of W. D. Ross, C. D. Broad, and their numerous fol-
lowers.[2] According to that theory, which remains popular in an
underground way, Davidson's contradiction-entailing premises
only appear to be so: correctly understood, they are covertly qualified as *"prima facie"* only. The full sense of "I ought to do it because
it will save a life, I ought not because it will be a lie" is *"Prima facie*
I ought to do it because it will save a life, and *prima facie* I ought
not, because it will be a lie." And that is equivalent to: "There is a
(moral) reason for doing it, namely that it would save a life, but
conceivably it may be overridden by a stronger reason; and likewise
there is a (moral) reason for not doing it, namely that it would be a
lie, but conceivably that too may be overridden by a stronger reason." Plainly, there is no contradiction here. What appeared to be
contradictory premises are now revealed as no more than statements of defeasible reasons, one for the action in question, and one
against it. According to neointuitionist theory, a moralist's task is
to determine which reason or set of reasons bearing on the question
whether or not something proposed is to be done outweighs and so
overrides those on the other side. There are no formal procedures
by which such tasks can be performed: they call for good judgment,
not for reasoning. Moral reflection is not a matter of practical reasoning, as it is commonly understood, but of surveying considerations intuitively perceived to be relevant, in the hope that none will
be overlooked, and then of intuitively gauging their relative weight.

This theoretical strategy, however, is suicidal. No theory of
moral reasoning can dispose of the objection that contradictions follow from premises it recognizes as true by denying that those premises really are premises. Nor is it easy to accept that reaching conclusions about what one is morally obliged to do is not a matter of
reasoning—that, as Davidson has satirically put it, the sum of our
moral wisdom concerning what to do in a given situation has the
form: there is something to be said for doing so and so, and something to be said against—and also for and against not doing it (35).[3]
It is true that from the fact that moral wisdom is reasoned and hence
has an analyzable structure, it does not follow that it is usefully formalizable. Legal reasoning is not usefully formalizable, and yet its
structure can be analyzed. Judges are perhaps not unknown whose
findings consist of listing considerations pro and con followed by
announcing the result of an intuitive weighing, but their reputations
are not high.

There is a class of moral theories in which moral reasoning is taken to resemble legal reasoning. Members of that class are sometimes described as "rationalist," in my opinion happily. Historically, the best known rationalist theories of morality are the theory of natural law that can be extracted from St. Thomas Aquinas's theological writings, and the theory expounded by Kant in the two volumes of his *Metaphysik der Sitten,* for which his *Grundlegung* broke the ground. Most rationalist theories of the present day are deeply indebted to Aquinas or Kant or both. The chief formal characteristics of such theories are five: (1) they rest on a few fundamental principles, sometimes one, which are advanced as true without exception; (2) each of those principles lays down some condition upon all human action as being required by practical reason; (3) those principles do not constitute a set of axioms, from which all the remaining moral precepts of the theory can be deduced; but, rather, (4) the remaining moral precepts are deduced from the fundamental principles by way of additional premises specifying further the conditions those principles lay down as required of all human action; and (5) both principles and additional premises are adopted on the basis of informal dialectical reasoning.

No set of formal characteristics, however, can capture what gives life to the rationalist approach to morality. When, as sometimes happens, a particular rationalist theory turns out to have implications that fall foul of dialectical considerations at least as strong as those on which it rests, either its principles or its additional premises must be revised. The fundamental methodological idea of rationalism is the nonformal idea that no revision of a premise may be ad hoc, merely intended to obviate an obnoxious implication; each must also turn out either to accord better with the dialectical considerations on the basis of which the unrevised premise was accepted, or to follow from a line of dialectical reasoning that is intrinsically superior.

A simple illustration may be found in Kant's *Lectures on Ethics.*[4] From the principle that no rational being is to be treated merely as a means, it is tempting to argue that, since giving to another information you believe to be false so far reduces him to a mere means to your purposes, it must be wrong to make any utterance to another that is contrary to your mind. Most moralists agree that an ancient example shows this to be a simple-minded error: namely, that of an armed would-be murderer who threatens an unarmed bystander with death unless he is informed of the whereabouts of his innocent quarry. How could it be wrong to give false information in such a case? But the new precept to which this example leads, namely that falsehood is wrong only in conditions of free communication where violence is neither done nor threatened to anybody concerned, is not ad hoc. It is derivable from the same principle as the unrevised one, but by way of an additional premise for which the dialectical grounds are stronger than those for the premise by which the rejected precept was derived. For the latter premise, that all falsehoods reduce those to whom they are told to mere means, is plainly indefensible when those to whom they are told have resorted to violence. It is, after all, a main theme in Kantian moral theory that not even the use of force to restrain those who use or threaten violence reduces them to mere means.

The problem of moral conflict is dismissed by rationalists as spurious, root and branch. In the following passage from the Introduction to his *Metaphysik der Sitten,* often quoted, but as a rule with astonishment or even incredulity, Kant gave classical expression to what is common to all of them:

> Because, however, duty and obligation are in general concepts that express the objective practical necessity of certain actions and because two mutually opposing rules cannot be necessary at the same time, then, if it is a duty to act according to one of them, it is not only not a duty but contrary to duty to act according to the other. It follows, therefore, that a collision of duties and obligations is inconceivable *(obligationes non colliduntur).* It may, however, happen that two grounds of obligation, one or the other of which is inadequate to bind as a duty *(rationes obligandi non obligantes),* are conjoined in a subject and in the rule that he prescribes to himself, and then one of the grounds is not a duty. When two such grounds are in conflict, practical philosophy does not say that the stronger obligation holds the upper hand *(fortior obligatio vincit),* but that the stronger ground binding to a duty holds the field *(fortior obligandi ratio vincit).*[5]

This passage is sometimes interpreted as anticipating neointuitionism by resolving some alleged conflicts of obligations into conflicts of grounds of obligation. But that is textually indefensible. In describing conflict between grounds of obligation, Kant avoided the metaphor for victorious struggle, namely, "holding the upper hand," which he employed in depicting what a putative conflict of obligations would be, and instead represented it as a conflict in which one ground 'holds the field', while the other, a mere *ratio obligandi non obligans,* simply vacates it, as being inadequate to bind as a duty.

What specifically has Kant in mind here? The distinction between perfect and imperfect duties in the *Grundlegung,* which is elaborated in *Metaphysik der Sitten,* supplies the answer.[6] Grounds of obligation of two kinds: those binding to perfect duties—duties to perform certain kinds of action other than those by which one executes the rational policies of self-culture or beneficence it is one's duty to form for oneself; and those binding to imperfect duties—or duties to execute such policies. Kant considered it obviously impossible that perfect duties or their grounds could ever be in collision. And he also held it to be impossible that imperfect duties could be in collision either with one another or with perfect duties: not obviously, but because of a condition on rational policies of self-culture or beneficence, namely that such policies must themselves be consistent and must not entail violating other duties, whether perfect or imperfect. In themselves, apart from this condition on rational policies, grounds of self-culture and beneficence can of course be in conflict with themselves, with one another, and with perfect duties: but so taken, they are for that very reason *rationes obligandi non obligantes* and, hence, inadequate to bind as duties.

Neither the rationalist nor the neointuitionist denial of the possibility of a genuine conflict of duties has been prominent in philosophical discussions of moral conflict in the past twenty years or so. Two innovations have captured most attention. One has been the elaboration, by John Rawls, of a contractarian version of Kant, according to which a system of morality is to be constructed by considering what precepts rational persons would agree to accept as binding, if the agreement were reached under conditions designed to secure a morally sound result. Consistency is obtained by providing that the precepts agreed upon be assigned a lexical order and that observance of those later in the order be conditional upon observance of those earlier in it. In his *Theory of Reasons for Action,* David A. J. Richards offers a specimen of such a system of lexically ordered precepts.[7] But it would be out of place, in a study of consistency in rationalist moral systems, to investigate this contractarian device; for if the rationalist approach to morality is sound, lexical ordering is either superfluous, because any morally necessary restriction on the scope of a precept must be deducible from the principles of an adequate system, or ad hoc, because it would exclude possible conflicts of precepts by imposing restrictions that are not thus deducible. The second innovation, however, is another matter. It is the acceptance, by a number of original and influential philosophers, of moral conflict as being with us much as the poor have been believed to be: either necessarily and always, or at least for a very long time.

Bernard Williams has succinctly expressed the idea underlying this innovation. Neointuitionist and rationalist moral theories, he contends, "do not do justice to the facts of regret and related considerations: basically because they eliminate from the scene the *ought* that is not acted upon."[8] It simply is not the case that, when one ground for acting is held to be stronger than another, the rejected ground is annihilated. "A man may feel regret because he has broken a promise in the course of acting (as he sincerely supposes) for the best; and his regret at having broken a promise must surely arise *via* a moral thought. . . . A tendency to feel regrets, particularly creative regrets, at having broken a promise even in the course of acting for the best might well be considered a reassuring sign that an agent took his promises seriously" (175) [p. 125 in this volume].

There is a familiar objection to this. Is it not simply inconsistent on the one hand to decide that it is for the best to act contrary to a certain ought and on the other to refuse to eliminate that ought from the scene? However, if that is inconsistent, it should be possible to show it.

In examples like Williams's, the conflict that compels the agent to act contrary to a certain ought is generated by his acceptance of three propositions:

(1) I ought to save this life.
(2) I ought to keep this promise.
(3) I cannot both save this life and keep this promise.

Yet, although accepting this triad, together with the belief that it is better to save this life than to keep this promise, leaves the agent no rational option but to break this promise, it does not follow that the triad, taken as it stands, is inconsistent. A contradiction is a proposition to the effect that something both is and is not the case. But that (1) an agent ought to act in a certain way, and that (2) he ought to act in a certain other way, do not together with the fact that (3) he cannot act in both ways, entail either that it is and is not the case that he ought to act in a certain way, or that it is and is not the case that he cannot act in both ways; and there is no other contradiction they can plausibly be supposed to entail. The fact that the triad generates a conflict does not show that it is inconsistent.

Why then do most moral philosophers reject as inconsistent any moral position that contains such triads? Because, as Williams has pointed out, they take moral positions generally to be committed to two presuppositions, one logical and one ethical *(ibid)*. The logical presupposition, which Williams calls "the agglomeration principle," is the generalization of the proposition that if it is the case that I ought to save this life and also the case that I ought to keep this promise, then it is the case that I ought both to save this life and to keep this promise. The ethical presupposition is expressed in the Kantian slogan "ought implies can," and can be formulated as "If you ought to do an action of a certain kind then you can do an action of that kind." Asserting the triad together with these presuppositions is certainly inconsistent; for from (1) and (2), together with the agglomeration principle, it follows that it is the case that I ought both to save this life and to keep this promise, whereas from (3) and the contrapositive of the Kantian slogan as I have formulated it, it follows that this is not the case.

Hence, if they are to avoid inconsistency, those who accept triads like the one we have been investigating because they agree with Williams that the moral ought that is not acted on cannot be eliminated from the scene, have no choice but to abandon either the agglomeration principle or the principle expressed in the slogan "ought implies can." Neither course is inviting.

Williams himself, followed by Bas C. van Fraassen, opts for the former. Because, as the facts of regret show, the ought that is not acted on is not eliminated even when it is for the best not to act on it, he argues that the picture of moral thinking according to which "there can be two things, each of which I ought to do and each of which I can do, but of which I cannot do both" is "more realistic" than one according to which there cannot, and that we should have no qualms at abandoning the less realistic of the two pictures (182) [p. 132 in this volume].

The contribution of van Fraassen has been to equip Williams's conclusion with a semantical interpretation. In standard systems of deontic logic, the "ought" operator is taken to be analogous to the necessity operator of modal logic; and in all standard systems of modal logic, if it is necessary that p and necessary that q, it is necessary that p and q. But van Fraassen has pointed out that there are command theories of morality according to which ought-statements of the form "It ought to be the case that p" are true if and only if some authority competent to issue moral commands has issued one (e.g., "Let it be brought about that p") that would not be fulfilled if it were not the case that p.[9] Such a command theory may exclude as ill-formed any putative command to do what cannot be done, so that "Let it be brought about that p and q," where it cannot be brought about that both are true, would not be a genuine command. Yet it would remain possible, on such a theory, that the competent authority should issue conflicting commands—commands each of which can be carried out, although all cannot be carried out together. "Let it be brought about that p" and "Let it be brought about that q" may each be genuine commands, even though "Let it be brought about that p and q" is not. Some divine-command theories of morality appear to be of this kind, for example, that of St. Gregory the Great, who held that human beings could be trapped by the devil into situations in which they were confronted with conflicting divine commands.[10]

Although most deontic logicians resist van Fraassen's proposal for amending the orthodox semantic foundations of their subject, the fundamental objection to adopting a semantics for ought-statements that would invalidate the agglomeration principle is ethical. Unless there is an alternative that is unknown to me, the only ethically plausible interpretation of moral systems that exclude agglomeration is van Fraassen's—that they are systems of commands by appropriate authorities. In such systems, provided that a command is in proper form and has been given by an appropriate authority, its moral validity is assured. However, although there are systems of this kind (divine-command systems being the most common), rationalists reject them as false in principle. Their position is that the only thing that can make a command a moral one is that it is imposed by practical reason. The commands God addresses to human beings as such are indeed morally binding; but that is because they express the infinite rationality of the divine essence, in which human rationality participates. To depict God as imposing on human beings commands contrary to practical reason is not only blasphemous but also absurd.

A moral theory according to which the ultimate moral authority gives commands that cannot be agglomerated conceives the moral universe as analogous to the U.S.S. Caine in Herman Wouk's *The Caine Mutiny,* and the moral authority as analogous to Captain Queeg. Even in the U.S.S. Caine, Kant's principle that ought implies can holds: a command to do the impossible is not a genuine command. However, two distinct commands, each of which can be obeyed, will each be legitimate, even if the situation is such that they cannot both be obeyed. True, at a court martial, it would be a perfect defense against a charge of disobedience that, having obeyed one of the conflicting commands, it thereby became impossible to obey the other. But, as long as Captain Queeg was the sole judge, one can imagine that this perfect defense would be rejected, as St. Gregory the Great imagined God rejecting a parallel defense to the charge of failing to obey conflicting divine commands. From the point of view of the lower deck, the universe of the U.S.S. Caine is absurd and unjust, but it is not inconsistent. And if the relation of human beings to moral authority is that of the lower deck to Captain Queeg, morality is absurd and nasty (we cannot say "unjust"!) but it is not inconsistent.

It is of the essence of rationalism to repudiate any conception of the relation of human beings to moral authority as analogous to that of the lower deck on the U.S.S. Caine to Captain Queeg. For according to rationalism, genuine moral requirements are requirements of practical reason on beings who themselves possess practical reason. Moral requirements may be asserted by authorities outside the individual moral agent, for example, the state (e.g., in criminal law). But no external requirement can be a moral requirement unless an adult of sound mind and normal education in a morally decent society, if he wishes to learn, can be brought to see its necessity for himself. Those who are under the moral law are therefore autonomous. It is as though Captain Queeg were unable to give any commands that did not belong to a set which, as a whole, the entire lower deck could be brought to acknowledge to be reasonable. In the moral universe, human beings stand to their moral authorities—the state, religious institutions, public opinion—not only as the lower deck stood to Captain Queeg, but as did the brother officers and equals who tried him at his court martial. And just as a court martial will judge a commander incompetent if he has made a practice of giving commands that must be ruled invalid when agglomerated, so autonomous moral agents will reject a moral system as ill constructed if there are situations to which its precepts apply, but in which their agglomeration would be invalid.

For this reason, the only semantics of ought-statements that appears to make sense of abandoning the agglomeration principle is ethically impossible. How, then, can situations occur at all in which moral obligations conflict? Among recent philosophers, the late E. J. Lemmon appears to have been the first to draw the inevitable conclusion: that such situations can occur is "a refutation of the principle that 'ought' implies 'can'."[11] But Lemmon's treatment was cavalier. The Kantian principle is too deeply entrenched in ethical thought to be dismissed outright. Not only rationalists are apt to be bewildered by what a moral obligation could be, if there should admittedly be no question of doing what one is morally obliged to do. Ruth Barcan Marcus, however, has recently proposed a revised version of the Kantian principle which promises both to make sense of moral obligations and to allow for the possibility of situations in which they conflict.[12]

The idea underlying Marcus's revision of the ought-implies-can principle originates in model theory. A set of meaningful sentences is consistent, she points out, provided only that it is possible for all the members of the set to be true: provided that there is a model, or possible world, in which they are all true. "Analogously," she goes on to say, "we can define a set of rules as consistent if there is some possible world in which they are all obeyable in all circumstances in *that* world" (128) [p. 194 in this volume]. If a moral system is consistent provided only that there is a possible world on which all its precepts are obeyable in all circumstances, the principle that ought implies can must be reformulated as two distinct principles, a first-order one and a second-order one. The first-order principle is that each and every nonagglomerated ought implies can: if you ought to do something, that something not being itself an agglomeration of things you ought to do, then you can do it. The second-order principle is that each agglomerated ought, if it cannot be obeyed in all circumstances in the actual world, must be such that one can try to bring about a world in which it can be obeyed in all circumstances, and one ought to try. But this principle is regulative only. As Marcus puts it:

> One ought to act in such a way that, if one ought to do x and one ought to do y, then one can do both x and y. But the second-order principle is regulative. This second-order 'ought' does *not* imply 'can.' There is no reason to suppose, this being the actual world, that we can, individually or collectively, however holy our wills or rational our strategies, succeed in wholly avoiding . . . conflict [of obligations]. It is not merely failure of will, or failure of reason [that produces such conflict]. It is the contingencies of this world (135) [p. 200 in this volume].

The semantics for the moral ought by which Marcus elucidates her revision of the ought-implies-can principle, like that by which van Fraassen gave sense to the negation of the agglomeration principle, is logically unobjectionable. And it has ethical attractions. Confronted with situations in which they cannot do what seems to be for the best unless they violate some moral principle they embrace, moralists have always been tempted to find fault with the contingencies of the world rather than with their moral thinking. When Cicero lamented to Atticus that Cato seemed to think that he lived in Plato's republic, and not in the cesspool of Romulus, he was objecting, not to Cato's principles, but to his rustic notion that principles were to be acted on.[13] Is morality not an ideal, after all? How can we be blamed for falling short of our ideal when a flawed world puts its attainment beyond our reach? However often the ideal we should like to be able to live up to is violated by the best we can actually do, it remains our ideal; we are remorseful at having done what we had to do, and we strive (without much hope, it is true) to bring about a world in which we shall no longer have to do it.

Like Cato, rationalists have always rejected this Ciceronian position. Morality, as they conceive it, is a system of precepts defining the limits imposed on human action by practical reason; and human action takes place in the actual world—in Cicero's cesspool of Romulus. But no human being knows completely either that part of the course of events constituting the actual world which bears upon his own actions or the causal principles that obtain in it. Hence, while an adequate rationalist theory of morality is not concerned with mere fantasy worlds, it must lay down what practical reason requires of human beings not only in the actual world, but in any possible world that, for all they know, may turn out to be actual. According to the moral tradition that has descended to us, through Christianity, from Greek philosophy and Jewish religion, rational beings act freely in a course of events in which everything except the actions of rational beings forms part of a lawful natural order.[14]

I do not think that Davidson, or Williams, or Marcus, or any of the phalanx of distinguished contemporary philosophers who, by maintaining the reality of moral conflict implicitly reject the possibility of an acceptable rationalist moral theory, would deny that such a theory would be desirable if it were possible. Why are they convinced that it is not? Why does it seem to them virtually beyond question that some moral conflicts, at least, are not illusions symptomatic of the failure of thought, but genuine discoveries?

To judge by the examples they offer, most critics of rationalism appear to follow Henry Sidgwick in holding that no rationalist the-

ory of the duty of keeping promises can fail to generate moral conflicts, both by itself (conflicts between promises) and in relation to
other parts of morality (conflicts between promise-keeping and
other duties). An example of the first kind is described by Marcus,
when she points out that I may "make two promises in all good faith
and reason that they will not conflict" and yet find that they do conflict "as a result of circumstances that were unpredictable and
beyond my control" (125) [p. 192 in this volume]—as when, having
a thousand dollars in my office safe and five hundred in my pocket,
and owing Smith and Jones five hundred dollars each, I promise
each that I will repay my debt at my office tomorrow, only to find
next day that I cannot do so, because overnight my safe has been
emptied by a burglar. An example of the second kind has already
been touched on in discussing the agglomeration principle. I have
promised to transact some business with an associate in his office at
noon; but on my way I pass two cars that have collided, and am
begged by paramedical officers to help them to treat the injured,
because otherwise at least one will die. In this unforeseeable contingency, I cannot both save the life of the injured person and keep my
promise.

   Although some rationalists have treated promising less com-
petently than others, most of them agree that no genuine conflict of
duties occurs in cases of either of these kinds, and many agree also
on the principles by which this can be shown. The fundamental
principle is that *it is morally wrong to make a promise unless you
can keep it and it is morally permissible for you to keep it.*[15]

   By itself, this principle would be self-defeating. With respect to
most promises that matter, promiser and promisee alike know that
unforeseeable contingencies can occur that would make it either
impossible or wrong for the promiser to keep his word. If, therefore,
it were wrong to make a promise unless one knew that no contingencies would occur that would make it either impossible or wrong
to keep it, promising would almost always be wrong. But it is not.
How is that possible? I submit that it is because most of us are willing to accept a promise provided that we trust the promiser and
understand him to have acceptable reason to believe that he can and
may do what he promises. True, we recognize that acceptable reasons may not be sufficient reasons: owing to unforeseeable contingencies it may turn out to be impossible or wrong to keep the promise. Since we accepted the promise on the understanding that things
might so turn out, should they do so we would not be entitled to

demand that it be kept, although, depending on circumstances, we might be entitled to amends for the breach. Hence we have a second principle: *most promises are made and accepted on the twofold condition that the promiser has acceptable reason to believe that he can and may do what he promises, and that if nevertheless it turns out that he either cannot or may not, the promisee will not be entitled to performance.* If the promiser fails to satisfy his part of this twofold condition, he does wrong in making the promise, and his consequent moral difficulties are his fault, not the fault of circumstances or of the moral system. If the promisee disregards the condition on what he is entitled to, and demands that the promiser keep his promise even though, through no fault of his own, he either cannot or may not keep it, the promiser may reject that demand as contrary to a condition on which his promise was accepted. This twofold condition on the collaborative act of making and accepting a promise, unlike internal conditions which vary from promise to promise, is virtually universal, and hence is normally unexpressed.

A difficulty, however, remains. It sometimes happens that, although a promise is given and accepted on the condition just described, the promisee would not consider acceptable the reason upon which the promiser believes that he can and may keep his word. This disagreement may not appear when the promise is made, because a promisee normally does not ask the promiser what his reason is for believing that he can and may keep his word. This difficulty is resolved by a third principle: that *it is wrong for a promiser to make a promise on any condition on which he does not believe the promisee to understand him to make it.* As William Whewell pointed out,[16] it cannot be seriously maintained, even by disgruntled promisees, that a promiser binds himself to do more than he believes the promisee, in accepting the promise, takes him to bind himself to do.

When I find myself in the fix that if I do something I will break my word to Lavinia and that if I do not I will break my word to Lolita, then according to the first of these principles, there is some reason to presume that, in view of my promise to Lolita, it was wrong to make to Lavinia the promise I did. But what if I had reason to believe that I could keep my promises to both? Then, according to the third principle, I must ask whether I believed that Lavinia, had she known of my promise to Lolita, would have thought my reason for believing that I could also keep my word to her to have been acceptable. If I did not, then I wrongly made Lavinia a promise on a condition on which I did not believe she understood me to have made it. On the other hand, if I did, then by the second principle, since there was a condition on my giving my word to Lavinia that has not been fulfilled, I am not bound to do what I promised her, although I may have amends to make for not doing it. It will be obvious how the three principles apply to the examples given of the two kinds of alleged conflict of duties to which promising gives rise.

Promising survives as an important institution, and not merely as an amiable ritual, because promiser and promisee can often be confident, whether from a shared culture or from personal intimacy, that they would agree about the acceptability or unacceptability of any reason that might be put to them for believing that a particular promise could be kept. You show that giving your word is a serious matter by your scrupulousness in ensuring that those to whom you give it understand the conditions on which you do so and in keeping it when those conditions are fulfilled, not by irrational guilt at breaking it when they are not.

I do not imply that before now any rationalist moralist has resolved in exactly this way conflicts of duties allegedly generated by the institution of promising. But I do contend that many rationalist theories contain the elements of this way of resolving them: in particular, the principle that it is wrong for anybody to make a promise unless he can keep it and it is morally permissible for him to keep it; and the related principle that besides the internal conditions on what is promised, there are normally other conditions, usually unexpressed, on the act of promising itself. In view of that, I further contend that even if a given rationalist theory fails to exclude the possibility of situations in which morally permissible promises could generate irresoluble conflicts of duties, it can nevertheless, along the lines I have proposed, be so revised as to exclude it.

Yet there is an objection of principle to the idea underlying this way of resolving conflicts of duties arising from promises. That idea is that no genuine conflict of duties can be generated by a promise it was wrong to make. But is that true? Even conceding that it is wrong to promise to do what you cannot do or what it would be wrong to do, can a promiser plead that wrong as justifying his breaking his word? Has not the promisee reason to complain: "You should have thought of that when you gave me your word, which I accepted in good faith on the conditions on which you gave it; but it does not follow, because you did wrong to give it, that you do not also do wrong in breaking it"? This seems unanswerable; and if it is, those who wrongly give their word may entangle themselves in genuine conflicts of duties.

The rationalist reply to this objection, which was first clearly worked out by St. Thomas Aquinas, concedes its point, but deprives it of force by a distinction.[17] The doctrine that there can be no genuine conflict of duties is ambiguous. If it means that all the precepts of a true moral system are obeyable in all situations to which it applies, then it is true. But taken in this sense, it does not imply that somebody who violates some of its precepts in some situations will be able either to obey all its other precepts in those situations or all its precepts in other situations. And cases of conflicts of duty arising from wrongful promises show that it is false if it is taken in a stronger sense in which it would imply that not even wrongdoing can give rise to situations in which the wrongdoer is entangled in a conflict of duties.

Aquinas classically expressed this point by distinguishing two kinds of conflict of duties, or, as he called it, two kinds of "perplexity": perplexity *simpliciter* and perplexity *secundum quid (ibid.).* A moral system allows perplexity (or conflict of duties) *simpliciter* if and only if situations to which it applies are possible, in which somebody would find himself able to obey one of its precepts only if he violated another, even though he had up to then obeyed all of them. For reasons already given, Aquinas held that any moral system that allows perplexity *simpliciter* must be inconsistent. By contrast, a system allows perplexity (or conflict of duties) *secundum quid* if and only if situations to which it applies are possible in which, as a result of violating one or more of its precepts, somebody would find that there is a precept he can obey only if he violates another. With regard to perplexity *secundum quid* Aquinas remarked that there is nothing logically wrong *(inconveniens)* with a moral system according to which a person in mortal sin can find himself perplexed.[18]

Since not all rationalists have perceived the distinction between conflicts of duty *simpliciter* and conflicts of duty *secundum quid,* not all have been in a position to draw the conclusion that a moral system may be consistent and yet allow conflicts of duty *secundum quid*—conflicts that themselves arise out of violations of duty. And, despite a note in G. H. von Wright's *Essay in Deontic Logic* that contains everything essential,[19] neither Aquinas's distinction nor its implication appears to be well known. It is therefore hard to estimate how far the prevalent opinion that rationalist systems of morality must be inconsistent because they generate moral conflicts proceeds from the error that systems that generate moral conflicts *secundum quid* must be inconsistent. However, once cases of conflict *secundum quid* are excluded as irrelevant to the question of consistency, plausible cases of moral conflict generated by sophisticated rationalist systems become hard to find.

Most of the remaining cases that appear in recent philosophical discussion have a common genus. And light is thrown on that genus by Marcus's claim to have derived, from the analogy of Buridan's ass, a principle that can be relied upon to generate moral conflict even within moral systems having only one fundamental principle, like Kant's. She writes:

> . . . it can be seen that the single-principle solution is mistaken. There is always the analogue of Buridan's ass. . . . The lives of identical twins are in jeopardy, and, through force of circumstances, I am in a position to save only one. Make the situation as symmetrical as you please. A single-principled framework is not necessarily unlike the code with qualifications or priority rule, in that it would appear that, however strong our wills or complete our knowledge, we might be faced with a moral choice in which there are no moral grounds for favoring doing x over y (125) [p. 192 in this volume].

I take it that Kant, and most rationalists, would object that the question which child is to be saved is not a moral question at all, if *ex hypothesi* the moral considerations are symmetrical; and that it seems to be a moral question only if it is falsely assumed that any practical question about what to do in a situation in which morality places restrictions on what we may do must be a moral question.

Where the lives of identical twins are in jeopardy and I can save one but only one, every serious rationalist moral system lays down that, whatever I do, I must save one of them. By postulating that the situation is symmetrical, Marcus herself implies that there are no grounds, moral or nonmoral, for saving either as opposed to the other. Why, then, does she not see that, as a practical question, Which am I to save? has no rational answer except "It does not matter," and as a moral question none except "There is no moral question"? Certainly there is no moral conflict: from the fact that I have a duty to save either a or b, it does not follow that I have a duty to save *a* and a duty to save *b*. Can it be seriously held that a fireman, who has rescued as many as he possibly could of a group trapped in a burning building, should blame himself for the deaths of those left behind, whose lives could have been saved only if he had not rescued some of those he did?

Yet although there is no moral conflict in cases of this kind— no conflict between duties—there is practical conflict. The fireman has reason to help, and wants to help, every one of those who beg him to; and he is torn when he can act on one such reason only if he does not act on another equally strong. This is one of those conflicts of grounds of obligation which Kant distinguished from genuine conflicts of obligation.[20] If there were only one person in the burning building, his need for help would hold the field as a ground binding to a duty *(Verpflichtungsgrund);* but since there are other grounds of the same force and it is impossible to act on them all, none of them holds the field as a ground binding to a duty, and the fireman's only duty or obligation is to act on as many of them as possible, it does not matter which. Still, although in this situation none of the various grounds binds to a duty, they remain grounds and they are in conflict. Practical conflict between considerations that have force, especially when in other circumstances one or another of them would give rise to a moral obligation, is readily mistaken for moral conflict.

The reason many philosophers at the present time are convinced that no acceptable moral theory can exclude the possibility of moral conflict *simpliciter* appears to be that, like Marcus, they assume that the question, What shall I do? in any situation in which moral considerations are relevant, must be a moral question. Davidson's remark, for example, "I do not believe that any version of the "single principle" solution [of the problem of moral conflict], once its implications are understood, can be accepted: principles, or reasons for acting, are irreducibly multiple" (34), would miss the point if the irreducibly multiple principles or reasons for acting were not moral. But rationalists insist that for the most part they are not.

If, as rationalists maintain, morality is the sum of those conditions on human action which are unconditionally required by practical reason, then for the most part moral considerations will not suffice to answer the question, What shall I do? They will in exceptional cases. For example, if I have made a binding promise to mail a letter by today's last collection, and half a minute before that collection am at the post office letter in hand, then moral considerations do suffice to answer the question, What shall I do in the next half minute? But such cases are not the rule. In Kant's rationalist theory, for example, of those human actions which are morally permissible, very few discharge perfect duties; the larger number, but still few, that discharge the imperfect duties of self-culture and beneficence allow room for inclination and, hence, are partly determined by nonmoral considerations; but most, including many of those we think about hardest and find it hardest to decide on (such as those raising questions like, Shall I rebuild my house on Malibu beach? Shall I abandon the doctoral program in Assyrian and apply to business school? and Shall I propose marriage to Bathsheba?) have little or nothing to do with morality.

It is true that, in many situations, the considerations I may have to weigh in answering the question, What shall I do? are irreducibly multiple: considerations of desire, convenience, affection, indignation, and courtesy—along with those of morality. The rationalist position is that, in most cases, moral considerations do not suffice to answer the question, What shall I do? What they do suffice to answer is the very different question, What conditions are imposed by practical reason on what I may do? Davidson's premise is true: the reasons that bear on what we are to do are irreducibly multiple. And his conclusion follows from it: in human praxis generally, conflicts of reasons for acting are inescapable. But, since not all practical principles are moral principles, practical conflicts are not necessarily moral conflicts; and rationalists maintain that they never are.

The argument of this paper is simply that the rationalist position that moral obligations never collide has not been shown to be false and that the prevalent impression that it has been springs from three sources: confusion of practical conflict generally with moral conflict; overlooking the distinction between moral conflict *simpliciter* and moral conflict *secundum quid;* and neglect of the casuistical resources of the various rationalist ethical traditions. I do not contend that any rationalist theory of morality yet produced is completely acceptable either morally or logically; but I do contend both that several such theories (among which I number those of Aquinas and Kant) are, in essence, serious options for moralists, and also that, if they should prove inconsistent, their inconsistency will turn out to result from corrigible blemishes rather than from radical incoherence.

Notes

The first version of this paper was presented to Donald Davidson's seminar at the University of Chicago in Winter 1981, the second was a Matchette lecture at the University of Wisconsin at Madison in November 1981, and substantially the present version was the Donald J. Lipkind Memorial Lecture at the University of Chicago in May 1982. For detailed criticism, I owe much to the audiences on those occasions; and also to members of colloquia at the branches of the University of California at Santa Cruz, Riverside, Irvine, and Los Angeles, at Texas A & M University, at the California Institute of Technology, and at Marquette University, Milwaukee, to all of which versions were read.

1. *Essays on Actions and Events* (New York: Oxford, 1980), p. 34. Parenthetical page references to Davidson are to this book.

2. Ross, *The Right and the Good* (New York: Oxford, 1930); Broad, *Five Types of Ethical Theory* (London: Routledge & Kegan Paul, 1930); for a recent endorsement, Robert Nozick, "Moral Complications and Moral Structures," *Natural Law Forum,* XIII (1968): 1–50.

3. I have reversed the order of Davidson's sentences in presenting what he says.

4. Louis Infeld, trans. (New York: Harper & Row, 1963), pp. 226–28.

5. Königsberg: 2nd ed., 1798, p. 24 (Ak. ed., vol. 6, p. 224) [pp. 39–40 in this volume; the translation is different]. John Ladd's translation is the basis of that given here, but I have altered it in three places. The German text corresponding to the fourth sentence of my English version is: "wenn zwei solcher Gründe einander widerstreiten, so sagt di praktische Philosophie nicht: dass die stärkere Verbindlichkeit die Oberhand behalte *(fortior obligatio vincit),* sondern der stärkere Verpflichtungsgrund behält den Platz *(fortior obligandi ratio vincit).*"

6. *Grundlegung zur Metaphysik der Sitten* (Riga: 2nd ed., 1786), p. 53n. (Ak. ed., volume 4, p. 421n.).

7. New York: Oxford, 1971.

8. "Ethical Consistency," in *Problems of the Self* (New York: Cambridge, 1977), p. 175 [p. 125 in this volume]. [This paper originally appeared in *Proceedings of the Aristotelian Society,* supp. vol. xxxix (1965).]

9. van Fraassen, "Values and the Heart's Command," *Journal of Philosophy,* LXX, 1 (Jan. 11, 1973): 15–19 [pp. 148–52 in this volume]. I use only van Fraassen's "initial" interpretation of ought, which clearly reveals his fundamental idea, and not his final and more sophisticated formulation, which enables him to deal with complexities foreign to the subject of the present paper.

10. St. Gregory's position is set out in Kenneth E. Kirk, *Conscience and Its Problems* (London: Longman, 1927), p. 322. See also my *The Theory of Morality* (Chicago: University Press, 1977), p. 144.

11. "Moral Dilemmas," *Philosophical Review,* LXXI 2 (April 1962): 139–58 [chapter 5 in this volume].

12. "Moral Dilemmas and Consistency," *Journal of Philosophy,* LXXVII, 3 (March 1980): 121–36, esp. 133–135 [pp. 199–200 in this volume]. Parenthetical page references to Marcus will be to this paper.

13. "ille optimo animo utens et summa fide nocet interdum rei publicae; dicit enim tanquam in Platonis πολιτέα, non tanquam in Romuli faece sententiam" (*Epistulae ad Atticum,* II, 1, 8).

14. By contrast, for example, with the morality of Hinduism; see my *Theory of Morality, op. cit.,* pp. 32–36, for a brief discussion.

15. In *A Theory of Reasons for Action* (New York: Oxford, 1971), pp. 164–65, David A. J. Richards has attempted to formulate the "generally accepted critical attitudes concerning the use . . . of expressions like 'I promise'" on which the existence of promising as a social institution depends. Most rationalists would endorse his formulation as approximately right.

16. *The Elements of Morality,* 4th ed. (Cambridge: Deighton Bell, 1864), section 280, p. 155.

17. *Summa Theologiae,* I-II, 6 *ad* 3; II-II, 62, 2; III, 64, 6 *ad* 3; *de Veritate,* 17, 4 *ad* 8.

18. *de Veritate, 17, 4 ad 8.*

19. Amsterdam: North Holland, 1968, p. 81 note 1. As far as I know, this is the first reference to Aquinas's distinction in print by a recent philosopher: von Wright himself acknowledges that P. T. Geach drew his attention to the relevant passages in Aquinas's writings.

20. See note 5 above.

Bibliography

Ackermann, Robert. 1968–69. "Consistency and Ethics." *Proceedings of the Aristotelian Society* 69: 73–86.

Aiken, Henry David. 1962. *Reason and Conduct: New Bearings in Moral Philosophy.* New York: Knopf.

Al-Hibri, Azizah. 1978. *Deontic Logic: A Comprehensive Appraisal and a New Proposal.* Washington D.C.: University Press of America.

Anderson, Alan Ross. 1958. "A Reduction of Deontic Logic to Alethic Modal Logic." *Mind* 67: 100–03.

Anderson, Lyle V. 1985. "Moral Dilemmas, Deliberation, and Choice." *Journal of Philosophy* 82: 139–62.

Aquinas, St. Thomas. 1952–54. *Truth.* Trans. Robert W. Mulligan, S.J., *et al.* 3 vols. Chicago: Henry Regnery Company.

———. 1964–75. *Summa Theologiae.* Trans. Thomas Gilby, O.P., *et al.* 60 vols. New York: McGraw-Hill.

Åqvist, Lennart. 1984. "Deontic Logic." *Handbook of Philosophical Logic.* Eds. D. Gabbay and F. Guenthner. 2 vols. Dordrecht: Reidel. 2: 605–714.

Aristotle. 1980. *The Nicomachean Ethics.* Trans. W.D. Ross, Rev. J.L. Ackrill, and J.O. Urmson. Oxford: Oxford University Press.

Atkinson, R.F. 1965. "Consistency in Ethics." *Proceedings of the Aristotelian Society* supp vol 39: 125–38.

291

———. 1969. *Conduct: An Introduction to Moral Philosophy.* London: MacMillan.

Aune, Bruce. 1979. *Kant's Theory of Morals.* Princeton: Princeton University Press.

Bambrough, Renford. 1979. *Moral Scepticism and Moral Knowledge.* London: Routledge.

———. 1984. "The Roots of Moral Reason." *See* Regis 39–51.

Barry, Brian. 1965. *Political Argument.* Atlantic Highlands, NJ: Humanities Press.

———. 1984. "Tragic Choices." *Ethics* 94: 303–18.

Baumrin, Bernard H., and Peter Lupu. 1984. "A Common Occurrence: Conflicting Duties." *Metaphilosophy* 15: 77–90.

Benn, S.I. 1983. "Private and Public Morality—Clean Living and Dirty Hands." *Public and Private in Social Life.* Eds. S.I. Benn and G.F. Gaus. New York: St. Martin's Press. 155–81.

———. 1984. "Persons and Values: Reasons in Conflict and Moral Disagreement." *Ethics* 95: 20–37.

Berlin, Isaiah. 1969. *Four Essays on Liberty.* Oxford: Oxford University Press.

———. 1978. *Concepts and Categories: Philosophical Essays.* New York: Viking Press.

Blum, Laurence. 1980. *Friendship, Altruism and Morality.* London: Routledge.

Bonevac, Daniel. 1983. "Chellas on Conditional Obligation." *Philosophical Studies* 44: 247–56.

Bradley, F.H. 1927. *Ethical Studies.* 2nd ed. Oxford: Oxford University Press.

Broad, C.D. 1930. *Five Types of Ethical Theory.* New York: Harcourt, Brace and Company.

Bronaugh, Richard. 1975. "Utilitarian Alternatives." *Ethics* 85: 175–78.

Calabresi, Guido, and Philip Bobbitt. 1978. *Tragic Choices.* New York: Norton.

Castañeda, Hector-Neri. 1966. "Imperatives, Oughts, and Moral Oughts." *Australasian Journal of Philosophy* 44: 275–300.

———. 1974. *The Structure of Morality.* Springfield, IL: Charles C. Thomas.

———. 1975. *Thinking and Doing: The Philosophical Foundations of Institutions.* Dordrecht: Reidel.

———. 1978. "Conflicts of Duties and Morality." *Philosophy and Phenomenological Research* 38: 564–74.

Cavell, Stanley. 1979. *The Claim of Reason: Wittgenstein, Skepticism, Morality, and Tragedy.* Oxford: Oxford University Press/ Clarendon.

Chisholm, Roderick M. 1963. "Contrary-to-Duty Imperatives and Deontic Logic." *Analysis* 24: 33–36.

————. 1964. "The Ethics of Requirement." *American Philosophical Quarterly* 1: 147–53.

————. 1979. "Practical Reason and the Logic of Requirement." *Practical Reason.* Ed. Stephan Körner. New Haven: Yale University Press. 1–17. Comments by G.E.M. Anscombe, J. Raz, and J.W.N. Watkins, along with a response by Chisholm, follow on 17–53. Reprinted in shorter form without comments in Raz, *Practical Reasoning* 118–27.

Conee, Earl. 1982. "Against Moral Dilemmas." *Philosophical Review* 91: 87–97.

Cox, A.A. 1978. "Castañeda's Theory of Morality." *Philosophy and Phenomenological Research* 38: 557–63.

Dancy, Jonathan. 1983. "Ethical Particularism and Morally Relevant Properties." *Mind* 92: 530–47.

Daniels, Charles B. 1975. *The Evaluation of Ethical Theories.* Philosophy in Canada: A Monograph Series 1. Halifax: Dalhousie University Press.

Davidson, Donald. 1970. "How is Weakness of Will Possible?" *See* Feinberg 93–113. Reprinted in *Essays on Actions and Events.* Oxford: Oxford University Press/Clarendon, 1980. 21–42.

De Sousa, Ronald B. 1974. "The Good and the True." *Mind* 83: 534–51.

Dewey, John, and James H. Tufts. 1932. *Ethics.* rev. ed. New York: Henry Holt.

Donagan, Alan. 1977. *The Theory of Morality.* Chicago: University of Chicago Press.

————. 1984. "Consistency in Rationalist Moral Systems." *Journal of Philosophy* 81: 291–309.

Dworkin, Ronald. 1975. "Hard Cases." *Harvard Law Review* 88: 1057–109. Reprinted in *Taking Rights Seriously.* Cambridge, MA: Harvard University Press, 1977. 81–130.

————. 1977. "No Right Answer?" *Law, Morality, and Society: Essays in Honour of H.L.A. Hart.* Eds. P.M.S. Hacker and J. Raz. Oxford: Oxford University Press/Clarendon. 58–84.

Feinberg, Joel. 1961. "Supererogation and Rules." *Ethics* 71: 276–88. Reprinted in *Doing and Deserving: Essays in the Theory of Responsibility.* Princeton: Princeton University Press, 1970. 3–24.

————, ed. 1970. *Moral Concepts.* Oxford: Oxford University Press.

Føllesdal, Dagfinn, and Risto Hilpinen. 1971. "Deontic Logic: An Introduction." *See* Hilpinen, *Deontic Logic* 1–35.

Foot, Philippa. 1983. "Moral Realism and Moral Dilemma." *Journal of Philosophy* 80: 379–98.

Frankena, William K. 1973. *Ethics.* 2nd ed. Englewood Cliffs, NJ: Prentice-Hall.

Gewirth, Alan. 1978. *Reason and Morality.* Chicago: University of Chicago Press.

————. 1984. "Replies to My Critics." *See* Regis 192–255.

Gowans, Christopher W. 1985. "Objectivism and Realism in the Sciences and Morality." *Proceedings of the American Catholic Philosophical Association* 59: 308–18.

Greenspan, Patricia S. 1983. "Moral Dilemmas and Guilt." *Philosophical Studies* 43: 117–25.

Gregor, Mary J. 1963. *Laws of Freedom: A Study of Kant's Method of Applying the Categorical Imperative in the Metaphysik der Sitten.* New York: Barnes & Noble.

Griffin, James. 1977. "Are There Incommensurable Values?" *Philosophy and Public Affairs* 7: 39–59.

————. 1982. "Modern Utilitarianism." *Revue Internationale de Philosophie* 36: 331–75.

Guttenplan, Samuel. 1979–80. "Moral Realism and Moral Dilemma." *Proceedings of the Aristotelian Society* 80: 61–80.

Hamblin, C.L. 1972. "Quandaries and the Logic of Rules." *Journal of Philosophical Logic* 1: 74–85.

Hampshire, Stuart. 1977. *Two Theories of Morality.* Oxford: Oxford University Press.

————. 1983. *Morality and Conflict.* Cambridge, MA: Harvard University Press.

Hare, R.M. 1952. *The Language of Morals.* Oxford: Oxford University Press/Clarendon.

————. 1963. *Freedom and Reason.* Oxford: Oxford University Press/Clarendon.

————. 1980. "Moral Conflicts." *See* McMurrin 1: 169–93.

————. 1981. *Moral Thinking: Its Levels, Method, and Point.* Oxford: Oxford University Press/Clarendon.

Harman, G. 1975a. "Reasons." *Crítica* 7: 3–18. Rpt. in Raz, *Practical Reasoning* 111–17.

————. 1975b. "Moral Relativism Defended." *Philosophical Review* 84: 3–22. Reprinted in *Relativism: Cognitive and Moral.* Eds. Michael Krausz and Jack W. Meiland. Notre Dame: University of Notre Dame Press, 1982. 189–204.

Harris, John. 1974. "Williams on Negative Responsibility and Integrity." *Philosophical Quarterly* 24: 265–73.

Harrison, Ross. 1979. "Ethical Consistency." *Rational Action: Studies in Philosophy and Social Science.* Ed. Ross Harrison. Cambridge: Cambridge University Press. 29–45.

Hartmann, Nicolai. 1932. *Ethics.* Trans. Stanton Coit. 3 vols. New York: Macmillan.

Hauerwas, Stanley, and Alasdair MacIntyre, eds. 1983. *Changing Perspectives in Moral Philosophy. Revisions 3.* Notre Dame: University of Notre Dame Press.

Hegel, G.W.F. 1975. *Aesthetics: Lectures on Fine Art.* Trans. T.M. Knox. 2 vols. Oxford: Oxford University Press/Clarendon.

————. 1977. *Phenomenology of Spirit.* Trans. A.V. Miller. Oxford: Oxford University Press/Clarendon.

Herman, Barbara. 1985. "The Practice of Moral Judgment." *Journal of Philosophy* 82: 414–36.

Hill, Thomas E., Jr. 1983. "Moral Purity and the Lesser Evil." *Monist* 66: 213–32.

Hilpinen, Risto, ed. 1971. *Deontic Logic: Introductory and Systematic Readings.* Dordrecht: Reidel.

————, ed. 1981. *New Studies in Deontic Logic.* Dordrecht: Reidel.

Hintikka, Jaakko. 1971. "Some Main Problems of Deontic Logic." *See* Hilpinen, *Deontic Logic* 59–104.

Hoag, Robert W. 1983. "Mill on Conflicting Moral Obligations." *Analysis* 43: 49–54.

Hook, Sidney. 1960. "Pragmatism and the Tragic Sense of Life." *Proceedings and Addresses of the American Philosophical Association* 33: 5–26. Reprinted in shorter form in *Pragmatism and the Tragic Sense of Life.* New York: Basic Books, 1974. 3–25.

Howard, Kenneth W. 1977. "Must Public Hands Be Dirty?" *Journal of Value Inquiry* 11: 29–40.

Hudson, W.D. 1970. *Modern Moral Philosophy.* Garden City, NY: Anchor-Doubleday.

Hughes, G.E., and M.J. Cresswell. 1972. *An Introduction to Modal Logic.* London: Methuen.

Jackson, Frank. 1985. "Internal Conflicts in Desires and Morals." *American Philosophical Quarterly* 22: 105–14.

Kanger, Stig. 1971. "New Foundations for Ethical Theory." *See* Hilpinen, *Deontic Logic* 36–58.

Kant, Immanuel. 1909. "On a Supposed Right to Tell Lies From Benevolent Motives." *Kant's Critique of Practical Reason and Other Works.* Trans. Thomas Kingsmill Abbott. London: Longmans. 361–65.

————. 1948. *Groundwork of the Metaphysic of Morals. The Moral Law.* Trans. H.J. Paton. London: Hutchinson. 53–123.

————. 1965. *The Metaphysical Elements of Justice: Part I of The Metaphysics of Morals.* Trans. John Ladd. Indianapolis: Liberal Arts-Bobbs.

————. 1971. *The Doctrine of Virtue: Part II of The Metaphysic of Morals.* Trans. Mary J. Gregor. Philadelphia: University of Pennsylvania Press.

Kekes, John. 1986. "Moral Intuition." *American Philosophical Quarterly* 23: 83–93.

Kolakowski, Leszek. 1968. "In Praise of Inconsistency." *Marxism and Beyond: On Historical Understanding and Individual Responsibility.* Trans. Jane Zielonko Peel. London: Pall Mall. 231–40.

Kolenda, Konstantin. 1975. "Moral Conflicts and Universalizability." *Philosophy* 50: 460–65.

Kolnai, Aurel. 1978. *Ethics, Value, and Reality*. Indianapolis, IN: Hackett.

Kuhn, Thomas. 1977. "Objectivity, Value Judgment, and Theory Choice." *The Essential Tension: Selected Studies in Scientific Tradition and Change*. Chicago: University of Chicago Press. 320–39.

Ladd, John. 1958. "Remarks on the Conflict of Obligations." *Journal of Philosophy* 55: 811–19.

———. 1982. "The Poverty of Absolutism." *Acta Philosophica Fennica* 34: 158–80.

———. Unpublished article. "Moral Dilemmas."

Lemmon, E. J. 1962. "Moral Dilemmas." *Philosophical Review* 70: 139–58.

———. 1965. "Deontic Logic and the Logic of Imperatives." *Logique et Analyse* 8: 39–71.

Lyons, David. 1978. "Mill's Theory of Justice." *Values and Morals*. Ed. A.I. Goldman and J. Kim. Dordrecht: Reidel. 1–20.

MacIntyre, Alasdair. 1957. "What Morality Is Not." *Philosophy* 32: 325–35.

———. 1981. *After Virtue: A Study in Moral Theory*. Notre Dame: University of Notre Dame Press.

MacLean, Anne. 1984. "What Morality Is." *Philosophy* 59: 21–37.

Mallock, David. 1967. "Moral Dilemmas and Moral Failure." *Australasian Journal of Philosophy* 45: 159–78.

Mally, Ernst. 1926. *Grundgesetze des Sollens, Elemente der Logik des Willens*. Graz: Leuschner. Reprinted in *Logische Schriften: Grosses Logikfragment-Grundgesetze des Sollens*. Eds. Karl Wolf and Paul Weingartner. Dordrecht: Reidel, 1971. 227–324.

Marcus, Ruth Barcan. 1980. "Moral Dilemmas and Consistency." *Journal of Philosophy* 77: 121–36.

Margolis, Joseph. 1967. "One Last Time: 'Ought' Implies 'Can'." *Personalist* 48: 33–41.

McCloskey, H.J. 1969. *Meta-Ethics and Normative Ethics*. The Hague: Nijhoff.

McConnell, Terrance Callihan. 1975. "Moral Dilemmas and Ethical Consistency." Dissertation. University of Minnesota.

———. 1976. "Moral Dilemmas and Requiring the Impossible." *Philosophical Studies* 29: 409–13.

———. 1978. "Moral Dilemmas and Consistency in Ethics." *Canadian Journal of Philosophy* 8: 269–87.

———. 1981a. "Moral Absolutism and the Problem of Hard Cases." *Journal of Religious Ethics:* 9: 286–97.

———. 1981b. "Utilitarianism and Conflict Resolution." *Logique et Analysis* 24: 245–57.

———. 1986. "More on Moral Dilemmas." *Journal of Philosophy* 82: 345–51.

———. Unpublished Article. "Interpersonal Moral Conflicts."

McDowell, John. 1979. "Virtue and Reason." *Monist* 62: 331–50.

McMurrin, Sterling M., ed. 1980. *The Tanner Lectures on Human Values.* 5 vols to date. Salt Lake City: University of Utah Press.

Melden, A.I. 1959. *Rights and Right Conduct.* Oxford: Blackwell.

Mill, John Stuart. 1884. *A System of Logic, Ratiocinative and Inductive.* New York: Harper and Brothers.

———. 1961. *Utilitarianism.* Reprinted in *The Utilitarians.* Garden City, NY: Doubleday & Company.

Morris, Michael K. 1985. "Moral Dilemmas and Forms of Moral Distress." Dissertation. University of Pittsburgh.

Nagel, Thomas. 1972. "War and Massacre." *Philosophy and Public Affairs* 1: 123–44. Reprinted in *Mortal Questions.* Cambridge: Cambridge University Press, 1979. 53–74.

———. 1979. "The Fragmentation of Value." *Mortal Questions.* Cambridge: Cambridge University Press. 128–41. An earlier version appeared in *Knowledge, Value and Belief.* Eds. H. Tristram Engelhardt, Jr. and Daniel Callahan. Hastings-on-Hudson, NY: Institute of Society, Ethics and the Life Sciences, 1977. 279–94.

———. 1980. "The Limits of Objectivity." *See* McMurrin 1: 75–139.

———. 1986. *The View From Nowhere.* New York: Oxford University Press.

Nell, Onora. 1975. *Acting on Principle: An Essay on Kantian Ethics.* New York: Columbia University Press.

Nowell-Smith, P.H. 1972–73 "Some Reflections on Utilitarianism." *Canadian Journal of Philosophy* 2: 417–31.

Nozick, Robert. 1968. "Moral Complications and Moral Structures." *Natural Law Forum* 13: 1–50.

———. 1981. *Philosophical Explanations.* Cambridge, MA: Belknap-Harvard University Press.

Nussbaum, Martha. 1983. "Flawed Crystals: James's *The Golden Bowl* and Literature as Moral Philosophy." *New Literary History* 15: 25–50. Comments by Patrick Gardiner, Richard Wollheim, and Hilary Putnam follow on 179–200. Nussbaum replies on 201–208.

———. 1985. "Aeschylus and Practical Conflict." *Ethics* 95: 233–67.

———. 1986. *The Fragility of Goodness: Luck and Ethics in Greek Tragedy and Philosophy.* Cambridge: Cambridge University Press.

Papanoutsos, E.P. 1963. "Moral Conflicts." Trans. John P. Anton. *Philosophy and Phenomenological Research* 24: 73–82.

Parfit, Derek. 1984. *Reasons and Persons.* Oxford: Oxford University Press/ Clarendon.

Phillips, D.Z., and H.S. Price. 1967. "Remorse Without Repudiation." *Analysis* 28: 18–20.

Phillips, D.Z., and H.O. Mounce. 1970. *Moral Practices.* New York: Schocken Books.

Phillips, D.Z. 1982. *Through a Darkening Glass: Philosophy, Literature, and Cultural Change.* Notre Dame: University of Notre Dame Press.

Pincoffs, Edmund. 1971. "Quandary Ethics." *Mind* 80: 552–71. Reprinted in Hauerwas and MacIntyre 92–112.

Plato. 1974. *The Republic.* Trans. Desmond Lee. 2nd. ed. Harmondsworth: Penguin.

Platts, Mark de Bretton. 1979. *Ways of Meaning: An Introduction to the Philosophy of Language.* London: Routledge.

Price, Richard. 1969. *A Review of the Principal Questions in Morals. British Moralists.* Ed. D.D. Raphael. 2 vols. Oxford: Oxford University Press/Clarendon. 2: 131–98.

Primorac, Igor. 1985. "Hare on Moral Conflicts." *Analysis* 45: 171–75.

Prior, A.N. 1954. "The Paradoxes of Derived Obligation." *Mind* 63: 64–65.

Putnam, Hilary. 1981. *Reason, Truth and History.* Cambridge: Cambridge University Press.

Quinn, Philip L. 1978. *Divine Commands and Moral Requirements.* Oxford: Oxford University Press/Clarendon.

———. 1979. "Divine Command Ethics: A Causal Theory." *Divine Command Morality: Historical and Contemporary Readings.* Ed. Janine Marie Idziak. New York: Edwin Mellen. 305–25.

———. 1986. "Moral Obligation, Religious Demand, and Practical Conflict." *Rationality, Religious Belief, and Moral Commitment.* Eds. Robert Audi and William I. Wainwright. Ithaca, NY: Cornell University Press. 195–212.

Rabinowicz, Wlodzimierz. 1978. "Utilitarianism and Conflicting Obligations." *Theoria* 44: 19–24.

Raphael, D.D. 1955. *Moral Judgment.* New York: Macmillan.

———. 1974–75. "The Standard of Morals." *Proceedings of the Aristotelian Society* 75: 1–12e.

———. 1984. "Rights and Conflicts." *See* Regis 84–95.

Rawls, John. 1971. *A Theory of Justice.* Cambridge, MA: Belknap-Harvard University Press.

Raz, Joseph. 1975. "Reasons for Action, Decisions and Norms." *Mind* 84: 481–99. Reprinted in part in Raz, *Practical Reasoning* 128–43.

———, ed. 1978. *Practical Reasoning.* Oxford: Oxford University Press.

———. 1978. Introduction. *See* Raz, *Practical Reasoning* 1–17.

Regis, Edward Jr., ed. 1984. *Gewirth's Ethical Rationalism: Critical Essays with a Reply by Alan Gewirth.* Chicago: University of Chicago Press.

Rescher, Nicholas. Forthcoming. "Does Ought Imply Can?" *Ethical Idealism.* Berkeley: University of California Press.

Richards, David A.J. 1971. *A Theory of Reasons for Action.* Oxford: Oxford University Press/Clarendon.

Ross, Alf. 1941. "Imperatives and Logic." *Theoria* 7: 53–71.

Ross, W.D. 1930. *The Right and the Good.* Oxford: Oxford University Press/Clarendon.

———. 1939. *The Foundations of Ethics.* Oxford: Oxford University Press/Clarendon.

Sartorius, Rolf. 1975. *Individual Conduct and Social Norms: A Utilitarian Account of Social Union and the Rule of Law.* Encino, CA: Dickenson.

Sartre, Jean-Paul. 1975. "Existentialism Is a Humanism." Trans. Philip Mairet. *Existentialism from Dostoevsky to Sartre.* Ed. Walter Kaufmann. rev. ed. New York: Meridian-New American. 345–69.

Sayre-McCord, Geoffrey. 1986. "Deontic Logic and the Priority of Moral Theory." *Noûs* 20, 179–97.

Scheler, Max. 1963. "On the Tragic." Trans. Bernard Stambler. *Tragedy: Modern Essays in Criticism.* Eds. Laurence Michel and Richard B. Sewall. Englewood Cliffs, NJ: Prentice-Hall. 27–44.

Schneewind, J.B. 1970. "Moral Knowledge and Moral Principles." *Knowledge and Necessity.* Royal Institute of Philosophy Lectures 3: 1968–69. New York: St. Martin's-Macmillan. 249–62. Reprinted in Hauerwas and MacIntyre 113–26.

Schon, Donald. 1958. "Ultimate Rules and the Rational Settlement of Ethical Conflicts." *Philosophy and Phenomenological Research* 19: 53–64.

Searle, John. 1978. "*Prima Facie* Obligations." *See* Raz, *Practical Reasoning* 81–90. Reprinted in longer form in *Philosophical Subjects: Essays Presented to P.F. Strawson.* Ed. Z. van Straaten. Oxford: Oxford University Press/Clarendon, 1980. 238–59.

Sen, Amartya. 1980–81. "Plural Utility." *Proceedings of the Aristotelian Society* 81: 193–215.

———. 1985. "Well-being, Agency, and Freedom: The Dewey Lectures 1984." *Journal of Philosophy* 82: 169–221.

Sidgwick, Henry. 1981. *The Method of Ethics.* 7th. ed. Indianapolis, IN: Hackett.

Singer, Marcus G. 1984. "Gewirth's Ethical Monism." *See* Regis 23–38.

Sinnott-Armstrong, Walter. 1982. "Moral Dilemmas." Dissertation. Yale University.

———. 1984. "'Ought' Conversationally Implies 'Can'." *Philosophical Review* 93: 249–61.

———. 1985. "Moral Dilemmas and Incomparability." *American Philosophical Quarterly* 22: 321–29.

———. Forthcoming. "Moral Dilemmas and 'Ought and Ought Not'." *Canadian Journal of Philosophy.*

———. Unpublished article. "Moral Realism and Moral Dilemmas."

Slote, Michael. 1985. "Utilitarianism, Moral Dilemmas, and Moral Cost." *American Philosophical Quarterly* 22: 161–68.

Smith, Holly M. 1986. "Moral Realism, Moral Conflict, and Compound Acts." *Journal of Philosophy* 83: 341–45.

Solomon, W. David. 1985. "Moral Realism and Moral Knowledge." *Proceedings of the American Catholic Philosophical Association* 59: 41–57.

Steiner, Hillel. 1973. "Moral Conflict and Prescriptivism." *Mind* 82: 586–91.

———. 1983. "Reason and Intuition in Ethics." *Ratio* 25: 59–68.

Stocker, Michael. Unpublished article. "Dirty Hands and Conflicts of Values and of Desires in Aristotle's Ethics."

Stubbs, Anne. 1981. "The Pros and Cons of Consequentialism." *Philosophy* 56: 497–516.

Swank, Casey. 1985. "Reasons, Dilemmas and the Logic of 'Ought'." *Analysis* 45: 111–16.

Tännsjö, Torbjörn. 1985. "Moral Conflict and Moral Realism." *Journal of Philosophy* 82: 113–17.

Taylor, Charles. 1982. "The Diversity of Goods." *Utilitarianism and Beyond.* Eds. Amartya Sen and Bernard Williams. Cambridge: Cambridge University Press. 129–44.

Trigg, Roger. 1971. "Moral Conflict." *Mind* 80: 41–55.

Urmson, J.O. 1974–75. "A Defense of Intuitionism." *Proceedings of the Aristotelian Society* 75: 111–19.

Vallentyne, Peter. Unpublished article. "Four Types of Moral Dilemmas."

———. Unpublished article. "Prohibition Dilemmas and Deontic Logic."

Van Fraassen, Bas C. 1972. "The Logic of Conditional Obligation." *Journal of Philosophical Logic* 1: 417–38.

———. 1973. "Values and the Heart's Command." *Journal of Philosophy* 70: 5–19.

Von Wright, Georg Henrik. 1951. "Deontic Logic." *Mind* 60: 1–15. Reprinted in *Logical Studies.* London: Routledge, 1957. 58–74.

———. 1971. "Deontic Logic and the Theory of Conditions." *See* Hilpinen, *Deontic Logic* 159–77.

———. 1983. "On the Logic of Norms and Actions." *Practical Reason.* Vol 1 of *Philosophical Papers.* Ithaca, NY: Cornell University Press. 100–29.

Walzer, Michael. 1973. "Political Action: The Problem of Dirty Hands." *Philosophy and Public Affairs* 2: 160–80.

Wertheimer, Roger. 1972. *The Significance of Sense: Meaning, Modality and Morality.* Ithaca, NY: Cornell University Press.

White, Alan R. 1975. *Modal Thinking.* Ithaca, NY: Cornell University Press.

Whiteley, C.H. 1952–53. "On Duties." *Proceedings of the Aristotelian Society* 53: 97–104. Reprinted in Feinberg, *Moral Concepts* 53–59.

Wiggins, David. 1975–76. "Deliberation and Practical Reason." *Proceedings of the Aristotelian Society* 76: 29–51. Reprinted in shorter form in Raz, *Practical Reasoning* 144–52. Reprinted in different form in *Essays on Aristotle's Ethics*. Ed. Amélie Oksenberg Rorty. Berkeley: University of California Press, 1980. 221–40.

————. 1976. "Truth, Invention, and the Meaning of Life." *Proceedings of the British Academy* 62: 331–78.

————. 1978–79. "Weakness of Will, Commensurability, and the Objects of Deliberation and Desire." *Proceedings of the Aristotelian Society* 79: 251–77. Reprinted in *Essays on Aristotle's Ethics*. Ed. Amélie Oksenberg Rorty. Berkeley: University of California Press. 1980. 241–65.

Williams, Bernard. 1963. "Imperative Inference." *Analysis* 23 supp: 30–36. Reprinted with an additional note in Williams, *Problems* 152–65.

————. 1965. "Ethical Consistency." *Proceedings of the Aristotelian Society* supp vol 39: 103–24. Reprinted in Williams, *Problems* 166–86.

————. 1966. "Consistency and Realism." *Proceedings of the Aristotelian Society* supp vol 40: 1–22. Reprinted with an additional note in Williams, *Problems* 187–206.

————. 1972. *Morality: An Introduction to Ethics*. New York: Harper & Row.

————. 1973a. *Problems of the Self: Philosophical Papers 1956–1972*. Cambridge: Cambridge University Press.

————. 1973b. "A Critique of Utilitarianism." *Utilitarianism For and Against*. J.J.C. Smart and Bernard Williams. London: Cambridge University Press. 75–150.

————. 1976a. "Moral Luck." *Proceedings of the Aristotelian Society* supp vol 50: 115–35. Reprinted in different form in Williams, *Moral Luck* 20–39.

————. 1976b. "Persons, Character and Morality." *The Identities of Persons*. Ed. Amélie Oksenberg Rorty. Berkeley: University of California Press. 197–216. Reprinted in Williams, *Moral Luck* 1–19.

————. 1978. Politics and Moral Character." *Public and Private Morality*. Ed. Stuart Hampshire. Cambridge: Cambridge University Press. 55–73. Reprinted in Williams, *Moral Luck* 54–70.

————. 1979. "Conflicts of Values." *The Idea of Freedom: Essays in Honour of Isaiah Berlin*. Ed. Alan Ryan. Oxford: Oxford University Press. 221–32. Reprinted in Williams, *Moral Luck* 71–82.

————. 1981a. "Moral Obligation and the Semantics of *Ought*." *Ethics: Foundations, Problems, and Applications*. Proceedings of the 5th International Wittgenstein Symposium in Kirchberg. Eds. E. Morscher and R. Stranzinger. Vienna: Holder-Pichler-Tempsky. 71–76. Reprinted in different form as "*Ought* and Moral Obligation" in Williams, *Moral Luck* 114–23.

————. 1981b. *Moral Luck: Philosophical Papers 1973–1980*. Cambridge: Cambridge University Press.

————. 1982. "Practical Necessity." *The Philosophical Frontiers of Chris-*

tian Theology: Essays Presented to D.M. MacKinnon. Eds. Brian
Hebblethwaite and Stewart Sutherland. Cambridge: Cambridge
University Press. 145–52. Reprinted in Williams, *Moral Luck*
124–31.

———. 1985. *Ethics and the Limits of Philosophy.* Cambridge, MA: Harvard University Press.

Winch, Peter. 1965. "The Universalizability of Moral Judgements." *Monist*
49: 196–214. Reprinted in *Ethics and Action.* London: Routledge,
1972. 151–70.

Zimmerman, Michael J. Forthcoming. "Remote Obligation." *American
Philosophical Quarterly.*

———. Unpublished Article. "Lapses and Dilemmas."